Fibre Optic Systems

Fibre Optic Systems

Pierre Halley

CNET, ESE, France

Translated by John C. C. Nelson,
University of Leeds, UK

JOHN WILEY & SONS

Chichester . New York . Brisbane . Toronto . Singapore

First published under the title Les Systèmes à Fibres Optiques
Editions Eyrolles
Paris, 1985

Copyright © 1987 by John Wiley & Sons Ltd.

Library of Congress Cataloging-in-Publication Data:

Halley, Pierre.
  Fibre optic systems.

  Translation of: Les systèmes à fibres optiques.
  Bibliography: p.
  Includes index
  1. Optical communications.   2. Fiber optics.
I. Title.
TK5103.59.H3513   1987   621.38′0414   86–32569
ISBN 0 471 91410 X

British Library Cataloguing in Publication Data:

Halley, Pierre
  Fibre optic systems.
  1. Optical communications   2. Fiber optics
  I. Title   II. Les systèmes à fibres
  optiques. *English*
  621.38′0414   TK5103.59
ISBN 0 471 91410 X

Phototypesetting by Thomson Press (India) Limited, New Delhi
Printed in Great Britain by Page Brothers Ltd. Norwich

Contents

Foreword

Throughout history, for armies and for empires, a critical factor has been the distance over which commands and information could be transmitted without excessive delay. It is therefore natural that it should be a military expert, General Ferrié, who, as is well known, has pioneered the development of radio telegraphy in France.

The essential role played by the French Navy after the Second World War in the renewal of the French electronics industry applied to electromagnetic transmission and detection is, perhaps, less well known. The simple definition and execution of technical programmes was extended by the transfer to industry and research organizations of a brilliant group of young naval officers whose sound basic training had been enriched by years of operating experience and long-distance navigation.

It was not difficult to find a common language between eminent elders such as Maurice Ponte and Yves Rocard, from the most fundamental research and teaching at the highest level, and those who had developed a taste for applications during the changing fortunes of war and the resistance.

The author of this book, Pierre Halley, distinguished himself in this group first and foremost with a career entirely devoted to the service of the State. He left the Navy in 1959 in order to join the National Centre for Telecommunication Studies (CNET), from which he retired in 1979 for a very active retirement with the grade of Honorary Commander and the titles of Chevalier of the Legion of Honour and Officer of the National Order of Merit. The same profound unity characterizes his scientific vocation which is entirely devoted to the problems of propagation of electromagnetic waves, a discipline of which he is a specialist of international reputation, as evidenced by his nomination in 1975 as chairman of the 'Electromagnetic propagation' committee of AGARD, the scientific and technical section of OTAN.

The study of electromagnetic propagation—which could give the illusion of being a very narrow specialization—is, in fact, the very heart of the problem of telecommunications and its conclusions determine the nature of systems and the technology of their components. It is essential to respond to the challenge of the convexity of the terrestrial globe which opposes the law of practically rectilinear propagation in free space (without using very long wavelengths which carry little information).

A large part of Pierre Halley's career, as head of the Ionospheric Prediction Service, aimed to make the best use of the partial response which arises from the providential but temperamental existence of naturally reflecting atmospheric layers. For wavelengths too short to benefit from these reflections the only possibility is to enclose the globe in a network of discontinuous paths defined by successive relay stations; natural terrain was first used, with difficulty, followed by artificial satellites, which are infinitely more effective.

Whatever developments can be foreseen for this latter technique, a possibility of saturation of free space is arising—the 'ether' if a term abandoned by physics can be used by way of illustration—in relation to an increasing demand for transmission of information. Guided propagation appears as a virtually unlimited possibility of remedying this situation, with the optical cable as the most promising method.

In a substantial discussion of his favourite subject the author addresses a vast area in which progress must be made and continued in order to bring about the considerable potential advantages of optical transmission. Topics range from spectacular improvements in the transparency of glass to the flexibility necessary for the modulation of the optical signal if full use is to be made of the information capacity permitted by very high carrier frequencies. Specialists in numerous related subjects, as well as system designers, will find complete information on the state of the art in this book. In view of the general and constant reference to fundamental scientific principles it will enable a much larger class of readers to be reminded of, and perhaps better understand, new applications for light as well as ideas which today form part of the general culture of the engineer.

Although concerned with a very specific application, the book covers vast areas of science and technology and is rich in specific information, clear commentary and enthusiastic but objective views; it deserves a huge success.

Pierre Contensou
General Engineer of the Armament
Member of the Academy of Sciences,
France

Preface

This book is concerned with fibre optic systems. It appears at a time when large national and international telecommunication networks which are in operation and under development throughout the world are starting to be equipped with such systems or sub-systems.

Consequently, it seemed essential to describe the most important activities of the French telecommunications administration in this area. Therefore Chapter 15 includes a description of the Biarritz experiment and submarine cables based on discussions with colleagues at CNET and notes, reports and papers which have been received. These descriptions are necessarily incomplete and, where personal opinions are presented, do not commit the telecommunications administration.

Pierre Halley
(Former pupil of the Ecole Navale, Honorary Commander; former pupil of
the Ecole Supérieure d'Electricité, chief engineer at CNET(retired); Chevalier of the Legion
of Honour, officer of the National Order of Merit, Military Cross.)

1 *General introduction*

1.1 GUIDING OF LIGHT

When light from a distant star reaches the vicinity of the Earth the rays are almost parallel and the detected optical power is virtually independent of distance. The reduction in power per unit additional distance travelled by the light, for example 1 km, is zero.

At a short distance from the source the propagation of light through free space is quite different. When it flows in all directions from a point source it is said to 'diverge'. Because of this divergence, the received power varies greatly as a function of distance. This power decreases as the distance increases because fewer rays enter the receiving device.*

Optical fibres have a physical structure which tends to suppress the natural divergence of the light, that is, to confine it to a closed guide so that the reduction in optical power as a function of distance becomes very small. Hence, the power detected by a receiver at the far end of the fibre varies very little as a function of its length.

A means must be found to cause light from an external source to enter the fibre or even to produce light directly within the fibre itself, and therefore better solutions to these problems of light production and guidance by tapes or fibres are increasingly being found.

The luminous power is reduced to half in travelling a distance of 100 m, 500 m, 6 km or more in different cases. If the optical power can be halved a second time in the course of the guided propagation, without making the system difficult to produce, double distances of 200 m, 1000 m and 12 km respectively, could be covered. It is possible, therefore, to transmit signals over distances greater than 50 km. This is tending towards 100 km and the practical limit is certainly greater.

It is clearly possible to interconnect successive segments by means of repeaters and thereby cover greater distances. Furthermore—and this is of considerable interest—the optical fibre can be bent into a curve and the light will follow the curvature of the guide with only a small loss. What is not possible for light

---

* Divergence of radiant energy is not a characteristic particular to light. It occurs for all other kinds of radiant waves for which energy is lost by the source and transferred into the available region of space.

1

in space, where it propagates in a straight line at a distance from very massive objects, is possible in a guide. The optical fibre can be wound round a cylinder to make a coil and guiding of the light will still be assured.

Several independently modulated sources of light can be fed into the same fibre and transmitted simultaneously on account of the different modulation. For example, four signals can be propagated in one direction and four in the opposite one within the same fibre by combining them on entry and separating them as they emerge. This arrangement is called 'multiplexing' of the light, and it can be unilateral or bilateral. The light from one optical fibre can be distributed among several others and can be switched between fibres at great speed.

Possible applications are very numerous and wide-ranging. The one which seems to be the most important, because it satisfies a substantial existing need, involves the transmission of signals.

1.2 THE OPTICAL FIBRE OF TODAY

With a nominal diameter of 0.125 mm the optical fibre offers an enormous capacity for the transmission of information. For a good-quality fibre this capacity is almost equal to that of a communications satellite. Of all the advantages of fibres, in comparison with other means, this is by far the most important. Unlike ordinary glass fibre, as used for reinforcement, a fibre of optical glass, which has the diameter of a hair, does not have significant mechanical strength. It will be very quickly damaged by the environment if it is not protected by an impervious coating and made into a cable with reinforcements able to withstand mechanical forces.

Of course, any number of fibres can be included in a cable; for example, six or 12 and up to 144. Consider as an example an American marine cable with a single fibre:

Diameter of the optical fibre: 0.0125 cm
Diameter of the cable with its reinforcement: 0.125 cm
Cable weight: 2.09 kg/km
Breaking load: 250.0 kg
Strain at breakage: 3.5%

The reinforcement consists of 9840 glass fibres embedded in epoxy resin.
Consider also a 52 km long single fibre microcable having:

A breaking strength of 53 kg;
A weight in sea water of 1.011 kg/km

This cable can be launched by rocket.

For the same diameter and the same breaking force as a metallic coaxial cable the capacity of a submarine optical cable will be 400 times greater. However, with its minute cross-section of 0.0123 mm^2 and, for a length of 1 km, its volume of 12.3 cm^3 of refined vitreous material, the optical communication fibre cannot transmit useful power to the other end as a source of energy. When

Table 1.1. Evolution of communication techniques

Decade	Technical breakthrough
1900	Radio (Guglielmo Marconi)
1910	The triode valve (Lee de Forest)
1920	The television iconoscope (Vladimir Zworykin)
1930	Waveguides (Schelkumoff, Chu and Barrow)
1940	Radar (Robert Watson-Watt)
1950	The transistor (John Bardeen, Walter Brattain, William Shockley)
1960	The laser (Townes)
1970	The optical fibre (Kao and Maurer)
1980	Non-linear optics?

the intensity of the light passes a certain threshold (when the electromagnetic energy per unit volume becomes very great) the previously transparent and passive material no longer ignores the presence of photons. Generally undesirable effects occur and the optical fibre is not a transmission line for energy.

At present, the optical fibre represents a spectacular breakthrough in the progress of technology, and it is likely that all its possible applications have not yet been envisaged (see Table 1.1).

1.3 BRIEF HISTORY

Not long ago one could not envisage a communications system which used light waves instead of waves or microwaves from the radio spectrum. The advantages of a system in the optical band compared with a system in the radio band were, however, well known; the quantity of information which can be carried by an electromagnetic wave increases in proportion to its frequency. Now microwave frequencies are from 10 000 to 100 000 times less than those of light waves. By using light, four or five orders of magnitude can be gained in the quantity of information transmitted by the system.

However, the practical difficulties in manufacturing an optical system were considerable. The first attempts, which with handsight seem a little absurd, concerned propagation which was directed, but in open space and not in a closed guide. With the invention of optical instruments, followed by intense sources of artificial light such as the electric arc, discovered by Davy in 1813, and the incandescent lamp, a new era in optical communication was established during the second half of the nineteenth century. A projector and long or short pulses of light separated by dark intervals, following the code which Samuel Finley Morse had established around 1837–8, were used. Hence it became possible, in clear weather, by day or night, to receive telegraphic messages as far as the horizon from the source and to send these messages a greater distance by semaphore relays or the electric telegraph. During the Second World War the German marines used the *Lichtsprecher*, a 'ship-to-ship' or 'ship-to-shore' system of telephony using an infra-red carrier emitted by a rotatable projector.

However, it had been known for a long time that it is possible to propagate light in a physical guide established between two points. At Murano, near Venice, glass workers have been pouring glass since the Renaissance. Hundreds of thousands of visitors have been able to see the little star of light at the free end of the stream, showing the guided propagation of light from the fire in the curved stream, without anyone showing a real interest in the phenomenon and imagining a practical use.

The first serious suggestion of guided light dates from 1910, when Hondros and Petrus Debye published a theoretical study on the guiding of waves by dielectrics in multiple layers of transparent material. Other studies were published in 1920 and 1930 which did not attract sufficient attention because they were concerned with guiding of microwaves, which made radar possible. It was only in the 1950s that van Heel, Hopkins and Kanapy developed the 'fibrescope', and it was Kanapy who achieved the first glass fibre clad in glass and who called it a 'fibre optic'.

1.4 THE FIBRESCOPE

This is a very useful device which consists of a bundle of numerous parallel fibres forming a cord which is thin enough to be supple and flexible. The end sections are plane and have configurations which can be superposed. Each point of an end section is connected to its opposite number at the other end by a fibre. Thus illumination in black and white, or colour, at one end section is reproduced exactly on the section at the other end. The input section can be provided with an objective which forms the image of an object on it. The output section can be provided with a suitable eyepiece to magnify the image.

The device makes it possible to see into corners and round obstacles. The fibrescope is used, for example, to examine welds, nozzles and combustion chambers inside jet aircraft engines which would be inaccessible for observation without it. It is used equally in medicine, the small diameter and flexibility of the bundle of optical fibres allowing examination of the oesophagus and digestive passages without using surgery or X-rays. A practical fibrescope has associated peripheral equipment which facilitates its use (mechanisms for lighting, positioning, etc.).

1.5 THE LASER

The discovery of the laser and semiconductor light emitters such as the light-emitting diode boosted interest in optical fibres during the 1960s. Until then a source of light appropriate to fibres was not available. The light from an incandescent lamp could not be fed efficiently into a fibre but only a hundred-thousandth of the optical power of such a source. The laser was much more promising. Optical fibres became usable since light could be fed into them—and what light! Very intense, coherent, monochromatic, presented on a very small surface and with a small angular aperture!

For the first time, one could hope to inject watts into an optical fibre. Everything would depend on the dimensions of the input cross-section of the fibre.

What was the situation in 1967? The light losses in a good-quality fibre were about 1000 dB/km (giving an enormous attenuation of 10^{100}). These losses limited use to lengths of some tens of metres.

1.6 THE DISCOVERY OF TRANSMISSION WINDOWS

The first major event in this field dates from 1970, when Kapron, Keck and Maurer of Corning Glass Works in the USA, announced a technical break-through—the fabrication of several hundreds of metres of glass fibre with a measured attenuation of 20 dB/km. Finally, it was known that some amorphous materials such as glass could have a high transparency and allow very good guidance of light. It had been ignored until then.

Certainly, it was known that light could be guided, with a very small loss, in a cylindrical rod made from non-dissipating dielectric material, but it was not known how far this theoretical model could be approached in practice. Since then, progress has been made with the progressive elimination of the causes of attenuation (Table 1.2).

Metallic traces and water which pollute the dielectric material have been isolated and eliminated. Water, which enters almost all bodies and dissolves them to a measurable extent, introduces harmful H^+ and particularly OH^- ions into the glass.

At present, silica glass fibres can be made with a very small OH content which have an attenuation of 0.25 dB/km at a wavelength of 1.3 μm, and even better can be foreseen. Optical cables can also be readily manufactured which have an attenuation of 0.5 dB/km even when the fibres have been spliced, formed

Table 1.2. Attenuation of light, in the visible and near infrared, for several transparent materials in decibels per kilometre travelled

Material	Attenuation (dB/km)
Window glass	10 000
Optical glass (spectacles, objectives)	300
Pure water at $\lambda_0 = 0.5\,\mu m$	90
Silica glass optical fibres:	
Corning (USA) 1970	20
1972	4
1973	2
Fujikura (Japan) at $\lambda_0 = 0.83\,\mu m$	1
$\lambda_0 = 1.10\,\mu m$	0.5
Several $\lambda_0 = 1.30\,\mu m$	< 0.5
Approximate present limit	0.25

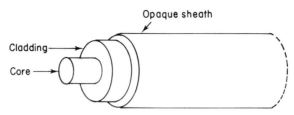

Fig. 1.1. Structure of an optical fibre: a central core and a peripheral transparent cladding, surrounded by a protective packaging

into a cable and laid. A length of 6 km is therefore possible for a loss of 3 dB, that is, half the optical energy. Distances between repeaters of several tens of kilometres are possible (Fig. 1.1).

1.7 THE DEVELOPMENT OF LIGHT GUIDES

The discovery of transparent optical materials caused intense research activity in order to discover new materials having even more interesting characteristics. This research started in the visible spectrum around 0.45 μm, advanced into the infra-red and progressed from there to longer wavelengths of 1.5–1.75 μm and beyond towards 2 or 3 μm or more.

The possibilities of Mendeleyev's table were explored to suggest mixtures of substances permitting vitrification with a suitable structure. The possibility was also studied of fabricating transparent tapes in the cold by mixing two dissolved

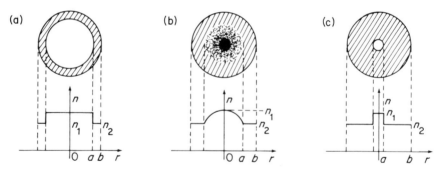

Fig. 1.2. Sections of the three principal types of optical fibre, associated with the configuration of their refractive index profile. The optical fibre is a thin rod of transparent material, and in general it is a cylinder of revolution about a central axis. (a) The cross-section of a step index fibre. In the centre, the core of radius a and refractive index n_1. Outside the core, the cladding of radius b and refractive index n_2. A fibre of this kind is multimode. (c) Represents the cross-section of a fibre of the same kind, whose core radius is small, so that the guide is monomode. (b) The cross-section of a graded index fibre. The refractive index n in the core decreases from the value n_1 to the value n_2 as the distance r from the axis of revolution increases from 0 to a. A fibre of this kind is practically always multimode

substances A and B and then evaporating the solvent after the reaction. The tape therefore consists of the transparent compound AB. The breakthrough was pursued and developed.

At present, the best compound is silica SiO_2, and the following are used:

(1) From 0.80 to 0.90 μm silica doped with boron and germanium (SiO_2 and B_2O_3 or GeO_2);
(2) From 1.00 to 1.60 μm silica doped principally with phosphorus (SiO_2 and P_2O_5)

More ordinary glasses are also used.

Numerous diameters of fibres were produced with a silica core clad in plastic. Finally, attempts were made to produce useful fibres with a plastic core clad in plastic (Fig. 1.2).

1.8 THE DIFFERENT TYPES OF OPTICAL FIBRES

Guiding the light changes its velocity as a function of the transverse co-ordinate of the guide, and a true physical discontinuity between the 'core' at the centre of the cylinder and the cladding can be obtained. This gives what is called a 'step index' (of refraction) fibre.

A progressive variation of index between the centre of the core and the cladding can also be obtained. This gives what is called a 'graded index' fibre. These two types of fibre result in different kinds of propagation of the light which are called 'modes' and which are separated from each other by diffraction.

Fibres of each of the above two types can be made which, for a given frequency v, have only one mode of propagation; a step index fibre having a core diameter sufficiently small for there to be only one guided mode is generally called a 'monomode' fibre. The diameter of the cladding is seven to ten times greater than that of the core.

Notice that a fibre of each of the three types (step index, graded index and monomode) always consists of a core, a homogeneous cladding where the light travels more quickly than in the core and a packaging which encloses, protects and isolates it.

1.9 THE PROPAGATION CHARACTERISTICS OF LIGHT IN AN OPTICAL FIBRE WHICH ARE ESSENTIAL TO ASSESS ITS QUALITIES

Optical fibres, even when sheathed in plastic, are considered to be linear optical guides. Vitreous silica is a very linear optical material. Its linearity disappears only at very high power fluxes, when the series expansion of the electric polarization can no longer be reduced to the first term:

$$\vec{P} = \chi_c \varepsilon_0 \vec{E}$$

Ordinary glasses are also very linear.

Linearity signifies that the output power W_2 from the fibre is proportional to the input power W_1. Doubling of W_1 produces a doubling of W_2. This is a very good approximation.

1.9.1 First essential datum: attenuation with distance

It is accepted that the optical power in the fibre decreases in a geometric series as the distance increases in an arithmetic series. In other words, the power is an exponential function which decreases with distance. The characteristic attenuation is specified in decibels per kilometre (dB/km) and is the global result of different loss processes:

(1) Absorption of light, intrinsically by the base material and extrinsically by the impurities;
(2) Scattering due to irregularities and, in the limit, to molecules. Other things being equal, it decreases by $1/\lambda^4$ as λ increases.
(3) Escape of light or failure of guidance (finite thickness of the cladding, curvature).

1.9.2 Second essential datum: channel capacity or bandwidth

In the transmission of digital signals one speaks of the channel capacity in bits per second (bit/s). In the transmission of analogue signals the unit is the bandwidth in Hertz (Hz). These two characteristics depend on the dispersion of the guide, and this is not a simple matter.

For monomode fibres and for multimode fibres over short distances the product of the bandwidth B of the fibre and its length L is constant: $B \cdot L = $ Constant. The characteristic $B \cdot L$ is therefore quoted in Megahertz·kilo-

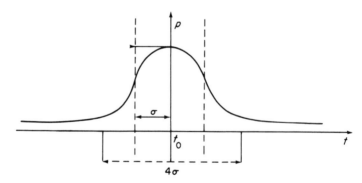

Fig. 1.3. The Gaussian function is extremely useful for representing a symmetrical pulse characterized by the time of the maximum t_0 and the standard deviation σ. The Gaussian function is:

$$\rho = \frac{1}{\sigma\sqrt{(2\pi)}} \exp\left(-\frac{\tau^2}{2\sigma^2}\right)$$

metre (MHz·km); it is often wrongly called the 'bandwidth.' Similarly, the characteristic $b \cdot L$ is quoted in (Megabit/second)·kilometre ((Mbit/s)·km).

If 2σ is the quadratic mean of the received impulse, in seconds, the 'golden rule' specifies that the channel capacity is $b = 1/4\sigma$. (Actually 96% of the total energy of a Gaussian impulse is received in a time interval of 4σ which is sufficient for a light impulse to be received and distinguished from 'darkness' of the same duration (Fig. 1.3).)

1.10 MODEL OF A GUIDED OPTICAL LINK

The optical fibre is a transmission channel for light whose use necessitates terminating components, the principal ones being suitably adapted emitters and receivers of light. Figure 1.4 shows the principles of the system: transmitter, transmission channel and receiver. Only the transmission channel is entirely optical. Couplers and separators of light, multiplexers and demultiplexers may be included. The light emitter is electronic-to-optical and the light receiver is optical-to-electronic.

1.11 MAXIMUM DISTANCE COVERED BY A SECTION OF AN OPTICAL LINK

The maximum distance which can be covered by the optical signal propagated in a guide, without the intervention of optical amplification, depends on:

(1) The minimum power level necessary for satisfactory reception. As this level is fixed, the maximum permissible attenuation of propagation in the guide

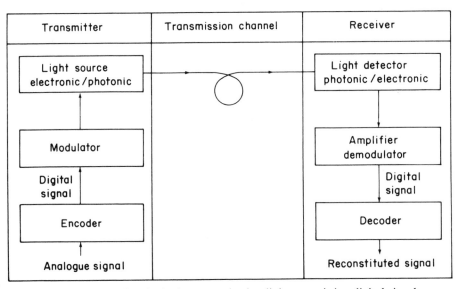

Fig. 1.4. Model of optical telecommunication link transmitting digital signals

is determined for a given source. A first limit, due to loss, occurs at a distance d_1 which depends on the attenuation characteristics of the guide.

(2) The bandwidth required by the user.

As this bandwidth is fixed, a second maximum permissible distance d_2 results which depends on the dispersion characteristics of the guide. The effective limit is the smaller of the two distance d_1 and d_2. This is the maximum distance between two electronic repeaters.

1.12 THE ADVANTAGES AND DISADVANTAGES OF OPTICAL FIBRES

1.12.1 Advantages

Fibres have characteristics which can be changed to suit requirements. It is possible, however, to state the most interesting aspects of performance and quality.

1.12.1.1 A very high digital data rate

The optical fibre is very well suited to the transmission of digital signals. The data rates of installed systems extend over more than three orders of magnitude and range from several Mbit/s to several Gbit/s (Fig. 1.5).

1.12.1.2 Very thin cables

Here, cable diameters are only millimetres. The space occupied by cable reels is very much reduced and the weight is small.

1.12.1.3 Coverage of long distances

The section covered without a repeater exceeds 40 km.

1.12.1.4 Immunity from interference and electromagnetic noise

Each light guide is closed. There is no interference and electromagnetic jamming is impossible. Lightning, radio transmissions and electromagnetic impulses from nuclear arms have no effect.

1.12.1.5 Good security

Interception by induction or simple contact is impossible. For interception, light must be extracted, and this can be detected.

1.12.1.6 Perfect safety

No electrocution, no short circuits and no earth leakage.

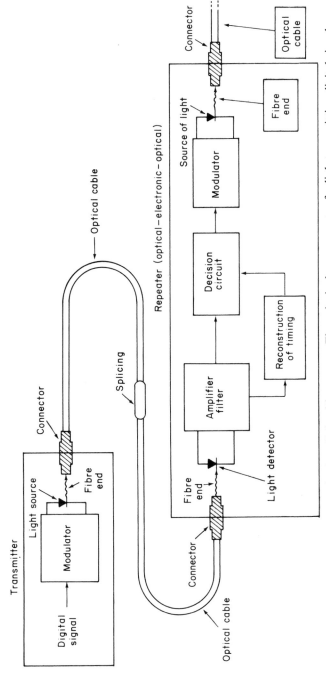

Fig. 1.5. Optical telecommunications link with repeater. The principal components of a link transmitting digital signals

1.12.1.7 The possibility of increasing capacity after commissioning

The capacity of an installed system can be doubled or tripled by multiplexing two or three (or more) light sources to an existing cable.

1.12.1.8 An attractive economic potential

A fibre carrying a hundred television channels would cost £50 per kilometre.

1.12.2 Disadvantages

1.12.2.1 The optical fibre cannot convey energy

When an intermediate repeater is necessary it must be powered separately; either locally or by a second (electrical) cable or by a combined optical and electrical cable.

1.12.2.2 Installation technicians must protect their eyes

The densities of optical energy emitted by the light source and by the extremity of the fibre, are sufficient to damage the retina permanently before the victim notices. It is essential to wear infra-red protecting glasses when working near operational equipment.

1.13 APPLICATIONS OF FIBRES AND OPTICAL CABLES

There are very many applications and there is much scope for development. One major group can immediately be distinguished by its economic importance—telecommunications applications.

1.13.1 Applications in national and international telecommunication networks

Optical fibres are particularly suited to point-to-point transmission of signals between communication centres. They permit analogue modulation since their power response is linear as well as digital modulation, particularly binary, with a light impulse representing the symbol 'one' and a dark space representing the symbol 'zero', for example.

High data rates allow the capacity of an installed network to be increased by a factor of between 10 and 50 by replacing metallic cables with optical ones in the existing conduits. This replacement also allows the number of repeaters to be reduced by a factor of around 20.

Metallic co-axial submarine cables already compete with communication satellites. Optical submarine cables in the main intercontinental communication routes will become the indispensable complement of satellites, and vice versa. A single type of digital signal allows transmission of telephone, viewphone, facsimile, computer data and television signals.

1.13.2 Applications in local networks

In addition to main networks, optical fibres allow the effective installation of local networks in an environment susceptible to electromagnetic noise. In this way, in a dense urban area the traffic police can observe major crossroads, bridges, tunnels, etc. on video in the police station using a network of remotely switchable optical fibres. With a similar network, six to twelve optically switchable cameras could observe for example, the stations and corridors of an underground station.

Other local networks could be installed in a large hospital, hotel, an entire estate, between a bank and its branches or between the local offices of a large daily newspaper.

1.13.3 Non-telecommunications applications

There are qualities and advantages of optical fibres other than those which relate to telecommunications; for example:

(1) Control and monitoring of electrical distribution without failure or risk during thunderstorms;
(2) Industrial instrumentation;
    (a) Measurement of pressure, from small sounds to explosions;
    (b) Measurement of electric field and of electron density. Gas or smoke detection (for example, in mines without introducing electric currents);
    (c) Temperature surveillance (in grain silos, for example);
    (d) Detection of nuclear radiation.
(3) The fibre optic gyroscope being developed for aircraft and spacecraft;
(4) Armaments; the optical fibre is light, not a burden and is insensitive to noise or electromagnetic interference.

This is only the beginning.

2　Light

The theoretical physicist has little difficulty in reconciling the different concepts which allow him or her to interpret and predict experimental facts concerning light. However, the communications engineer or technician can experience problems

From a practical point of view, the specialist must choose the approach which best suits his or her problem from the following three:

(1) *Light consists of photons.* There is an exchange of energy—emission, absorption and collision (light-emitting diodes, solid state lasers and photodetectors).
(2) *Light is an electromagnetic wave.* The medium is not dissipative. There is no exchange of energy, only propagation, interference and diffraction (monomode light guides, cavity resonances and diffraction gratings).
(3) *Light consists of rays.* The wavelength is infinitely small. The principles of geometrical optics apply (multimode guides, graded index guides and lenses).

(1) Light consists of photons

This is a propagation of energy which has electromagnetic properties and is emitted and captured by particles having an electric charge. In most cases this energy is simply exchanged between neighbouring particles, but it can find a space and propagate to a great distance. It cannot be detected without being at least partially absorbed.

Our eyes, which are sensitive to energy of this kind when it belongs to the 'visible' part of the spectrum, see objects which emit or scatter it and attribute a colour to them. Furthermore, we instinctively understand that light propagates in straight lines, in a medium which surrounds us and is generally homogeneous and transparent, at least for a short distance.

However the word 'light' is particular to optics, while electromagnetic interaction covers the whole spectrum from the lowest to the highest frequencies, passing through the radio spectrum, the infra-red spectrum, the visible band, the ultra-violet spectrum and X- and gamma-rays. It is therefore essential to speak of 'electromagnetic interaction'. This consists of one of four material interactions in which the vehicle or intermediate particle, which allows interaction at a distance, is the photon, the one and only particle in this interaction.

The wave theory of light was established during the nineteenth century under the influence of Augustin Fresnel and James Maxwell, and allowed light to be distinguished by its frequency.

However, the emission of light remained unexplained. In order to explain black-body radiation (the full radiator), which had been established experimentally, Max Planck proposed that it occurred discontinuously, in quanta, the energy of a quantum being proportional to frequency:

$$E = hv \tag{2.1}$$

where E is the energy, v the frequency of the theoretical oscillation and h is a constant whose value is established from experimental data:

$$h = 6.626 \, 10^{-34} \, (\text{J.s}) \tag{2.2}$$

Then Henri Poincaré showed that the discontinuous emission theory of light was absolutely necessary to obtain Planck's experimental formula. Two other phenomena, the photoelectric and the Compton effects, have shown the existence of particles of luminous energy.

Finally, if an electron (a negative particle) meets a positron (a positive particle) both electrically charged particles disappear and two photons are created. In the inverse operation two photons create a pair of negative and positive particles. All these changes observe the conservation of electrical charge, motion and energy.

The existence of the photon was no longer in doubt, and it has the following physical properties:

(1) An intrinsic mass of zero. Photons travel at the limiting velocity:

$$c \simeq 299 \, 792 \, 10^3 \, (\text{m/s}) \tag{2.3}$$

in free space.

(2) A zero electrical charge.

(3) A kinetic moment of spin $s = 1$, in units of $h/2\pi$. It is its own anti-particle.

All photons have these intrinsic characteristics and a photon or a group of identical photons can be distinguished by an additional extrinsic characteristic such as:

The frequency v

The energy $E = hv$

The wavelength $\lambda_0 = c/v$ (in free space) $\tag{2.4}$

The relativistic mass $m = hv/c^2$ $(E = mc^2)$ $\tag{2.5}$

These four characteristics are equivalent.

The photon, like all intermediate particles of interactions, does not obey Pauli's exclusion principle; that is, two photons can be in the same state which

16

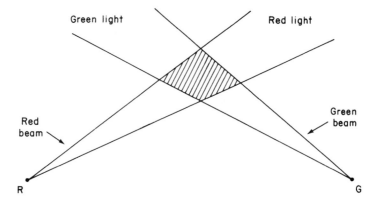

Fig. 2.1. Two beams of light which cross in air

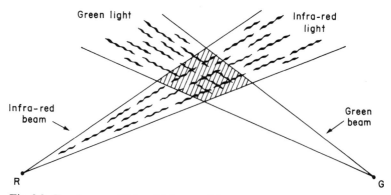

Fig. 2.2. Two beams of light which cross. The very powerful red beam heats the air and causes turbulence

is determined by the spatial co-ordinates and spin. Photons can be superposed. In contrast, electrons whose spin is $s = \frac{1}{2}$ obey the exclusion principle and cannot be in the same quantum state.*

The consequences of these facts are considerable. Two beams of electrons cannot cross without interacting; in contrast, two light beams can cross without difficulty.

Consider two light beams which cross in a non-dissipative (better still, totally transparent) atmosphere—a beam of red light from source R and a beam of green light from source G. After they have crossed, one finds separate beams of green and red light (Fig. 2.1). The red and green photons have crossed without their colour being changed, and nothing has happened in the common volume. One photon cannot act upon another except through an intermediate material, that is, by the operation of two interactions: here, for example, red

---

* One case (that is, a state defined by spatial co-ordinates) is saturated if it contains two electrons with opposite spins.

photon/material then material/green photon. If the air contains water vapour and if the beam R is infra-red and sufficiently intense, it will transfer energy into the dissipative air and cause turbulence by heating. This turbulence will partially destroy the coherence of the green beam within the common volume and irregularities will be seen in the green beam after it emerges (Fig. 2.2).

2.1 TRANSPARENCY AND OPACITY

Optical transparency is the property of a physical medium, such as free space, by which a photon of frequency v belonging to a luminous beam can propagate there without exchanging energy with the medium while still belonging to the beam. If the transparency is perfect all photons of frequency v will enjoy this property. A sheet of still water or a plate of glass with flat faces can be transparent for certain bands of frequencies.

A medium is non-dissipative when the photons propagate there without losing energy. That does not mean that a beam of light which propagates there will conserve its coherence. In the second example cited above, the common volume of air is non-dissipative for green light but the coherence of the beam is not conserved.

Opacity is the inverse of transparency and is caused either by the inhomogeneity of the propagating medium which destroys the coherence of the beam by causing scattering (for example, the light from car headlights is scattered by fog) or by absorption, which consumes optical energy and eventually transforms it to heat.

The materials in our environment are, in general, mixtures containing impurities, where scattering and absorption are combined. However, in order to obtain transparency it is essential to distinguish between scattering and absorption. They do not have the same causes and, consequently, not the same remedies.

Altogether, the inhomogeneities of a light-propagating medium and, particularly, the heterogeneities are the principal cause of opacity, since they cause it even in the absence of absorption. For example, snow composed of pure air and crystals of pure water is opaque, so are powdered glass and milk.

Only very homogeneous substances with very few absorbants are transparent. It can be stated that:

(1) *Absorption* refers to the loss processes by which a photon of frequency v disappears. It is the result of photon–material interactions inverse to the material–photon interactions of emission.
(2) *Simple scattering* refers to the loss processes by which a photon of frequency v is deviated and leaves the beam to which it belongs. This scattering is linear; that is, the mean number of photons scattered during unit time, in a small element of volume, is proportional to the mean number of incident photons belonging to the beam: $N\%$ of the photons of frequency v are scattered.

(3) *Scattering with exchange of energy* refers to the loss processes by interaction of three particles: a photon v, a phonon and a photon v'. The photon v disappears, but a photon of frequency v' greater or less than v appears, while an acoustic particle (the phonon) disappears or appears. This type of scattering is not linear.

2.2 RAYLEIGH SCATTERING

Material can be considered to be homogeneous only on a certain scale. A gas or a liquid can be considered as a typically amorphous mixture even on a molecular scale. Disordered solid material, such as glass, can present structural irregularities of small dimensions containing several tens to several hundreds of thousands of molecules.

In most cases it is possible to estimate the dimensions of the average irregularity. When the wavelength of the light is much greater than the mean dimension of the irregularity in the material in which it is propagated the resultant scattering is of the Rayleigh type, named after an English physicist (Lord Rayleigh, Nobel Prize winner, 1904) who explained it in 1900. If the molecule has a dimension of 3 Å and if the wavelength in the material is at least 10 times greater ($\lambda > 30$ Å), the scattering is linear and inversely proportional to the fourth power of the wavelength of the photon:

$$N\% = \frac{\text{Constant}}{\lambda^4} \tag{2.6}$$

If 100 photons are present, on average, N are scattered. If the wavelength is doubled, the mean number of photons scattered is divided by 16.

In consequence, Rayleigh scattering is not an important phenomenon except at sufficiently short wavelengths. We can observe it with the naked eye each day, since it is this which gives the sky its familiar appearance. If the Earth was surrounded by an absolutely homogeneous continuous medium the sky would be black and we would see the stars in the middle of the day at the same time as the Sun. This is not the case; the Earth is surrounded by a principally gaseous atmosphere, which contains not only liquids and solids in suspension but also molecules capable of scattering visible light.

Professor Alfred Kastler (Nobel Prize winner 1966) recalled recently that Goethe, the German writer, had noted the properties of a cloudy medium crossed by a beam of white light in his treatise on colours; the beam became reddish but appeared to be of a dark blue colour when viewed laterally within the medium against a black background. Goethe had combined his observations of the colours of the setting Sun and the blue light of the daily sky.

2.3 RAMAN AND BRILLOUIN SCATTERING

Chandrasekhara Raman, the Indian physicist and Nobel Prize winner (1930) and Léon Brillouin, the French physicist, discovered that these scatterings, which are two aspects of the same phenomenon, are non-linear. When a photon v collides with a molecule the impact is generally elastic and the scattered photon has the same energy and consequently the same frequency as the incident photon; this is Rayleigh scattering. However there can be an exchange of energy by inelastic collision and the molecule, which vibrates, gains energy from, or loses energy to, the incident photon.

This effect is very minor when the number of photons per unit volume is small. It can become major at high light intensities and stimulated Raman scattering and Brillouin back-scattering can occur in an optical fibre with solid or liquid core.

(2) Light is a wave

It is impossible to forego wave theory, which explains diffraction and interference so simply. This theory explains how light added to light can produce darkness, and completely explains diffraction, due to the work of Augustin Fresnel. Its consequences, which are very paradoxical to an inexperienced mind, are perfectly verified by experiment to a degree of precision which is greatly increased by the techniques used. These experiments have been repeated millions of times.

Only electromagnetic theory, which is based on the assumption of two independent vector equations proposed by James Clerk Maxwell, has enabled explanation of the relationship between light and radio waves as verified by experiment. Electromagnetic theory is without doubt applicable to optics.

2.4 WHAT IS AN ELECTROMAGNETIC WAVE?

A wave is a vibratory phenomenon which propagates (or may, on occasion, be stationary). Each vibratory phenomenon is primarily conditioned by its frequency. Only the wavelength and the group propagation velocity are directly accessible for experiment.

Light is a transverse electromagnetic wave. Action at a distance is produced by a field which propagates and appears as an electric or magnetic force; an 'object' allows this effect to occur in the region of space where the field prevails.

The wave can be represented by a surface which deforms during movement as a function of time. At each point on this wave surface the direction of propagation is normal to the surface.

The existence of these electromagnetic waves leads to consideration of free space, that is, empty space, as a physical medium which is characterized by three fundamental constants, related by:

$$\varepsilon_0 \mu_0 c^2 = 1 \tag{2.7}$$

which expresses the velocity c of the light wave as a function of two other constants: the electical permittivity of free space ε_0 and the magnetic permeability of free space μ_0. From a macroscopic point of view, and in the context of linear electromagnetism with which we are concerned, the surrounding materials will be themselves characterized by two coefficients and two only, ε and μ and by the velocity v of the light wave, which are related by:

$$\varepsilon\mu v^2 = 1 \qquad (2.8)$$

If the coefficients ε and μ are independent of the frequency v of the light, these are constant and v is a constant.

If all effects of diamagnetism, paramagnetism and, above all, ferromagnetism can be eliminated (and this will be assumed) we can put $\mu = \mu_0$ and write:

$$\varepsilon\mu_0 v^2 = 1 \qquad (2.9)$$

The velocity of light in the medium considered will be:

$$v = \frac{1}{\sqrt{(\varepsilon\mu_0)}} \qquad (2.10)$$

(In the units of the international system (SI) $\mu_0 = 4\pi.10^{-7}\,(\text{H.m}^{-1})$.)

2.4.1 The group velocity

If, in contrast, the coefficient ε is not constant, but $\varepsilon(v)$, a function of frequency, only the phase velocity v of the wave surface mentioned above will be given by:

$$v = \frac{1}{\sqrt{(\varepsilon(v)\cdot\mu_0)}} \qquad (2.11)$$

A group of waves, having a small spread in frequency around v, will show a 'modulation' which propagates with the group velocity u, on the 'carrier' of frequency v (see a following note).

Louis de Broglie's relation, which unites this velocity and the phase velocity v, is:

$$\frac{1}{u} = \frac{1}{v}\left(1 - \frac{v}{v^2}\frac{dv}{dv}\right) \qquad (2.12)$$

A medium in which the velocity v is independent of frequency is called 'non-dispersive'. In such a medium there is no particular group velocity at a frequency $u(v)$; the velocity of the light is simply v for all frequencies ($u = v$).

Unfortunately physical media are naturally dispersive and, in many cases, a non-dispersive model cannot be assumed. It is necessary to consider two velocities: that of the wave surface v at the point M where the local electrical permittivity is ε and that of the group of waves which is u.

(*NB*: Two plane waves of adjacent frequencies $v - dv$ and $v + dv$ propagate in the same direction with velocities $v - dv$ and $v + dv$, respectively. At a point

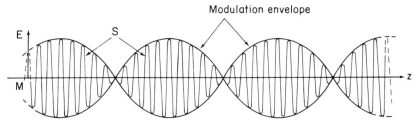

Fig. 2.3. The carrier propagates with velocity v. In the figure, the modulation envelope at the low frequency dv, which propagates at the group velocity u, is shown

M, the combination of these two waves is expressed by the product:

$$S = 2A \sin\left[2\pi v\left(t - \frac{z}{v} \right) \right] \cos\left[2\pi dv\left(t - \frac{z}{u} \right) \right] \qquad (2.13)$$

The first term, the sine, represents a wave of frequency v propagating with velocity v. The second term, the cosine, represents a modulation of frequency dv, propagating with velocity u. (Fig. 2.3))

2.5 THE REFRACTIVE INDEX

For a given medium and frequency values are defined relative to free space such that

$$\frac{\varepsilon}{\varepsilon_0} = \varepsilon_r \text{ (called 'relative electric permittivity')} \qquad (2.14)$$

$$\frac{c}{v} = n \begin{array}{l} \text{(called 'absolute refractive index')} \\ \text{(the inverse of the relative velocity)} \end{array} \qquad (2.15)$$

It can be seen that:

$$\varepsilon_r = n^2 \qquad (2.16)$$

2.6 THE TRANSVERSE FORM OF THE FIELD

Let us return to a representation of a wave surface which deforms as it moves and assume that at an adjacent point M, in the local homogeneous medium (that is, having the same properties at all points), the phase velocity is v. We can represent this velocity by a vector at M normal to the wave surface Σ (Fig. 2.4).

From Maxwell's equations it can be shown that the magnetic field vector $\vec{H}$ is always perpendicular to $\vec{v}$ and that the electric field vector $\vec{E}$ is always perpendicular to $\vec{H}$.

When in proximity to the point M, the medium is not only homogeneous but also isotropic (which implies that it has the same properties in all directions at every point); the vector $\vec{E}$ is also always perpendicular to $\vec{v}$ and the wave,

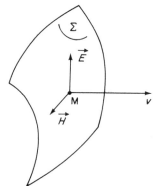

Fig. 2.4. Representation of a wave surface. Σ is the surface which the light, of frequency v occupies at time t having left the point source at time zero. For a homogeneous isotropic medium, the surfaces Σ are spherical

characterized by the two vectors $\vec{E}$ and $\vec{H}$ and contained in the plane tangential to Σ is of *transverse form*.

2.7 DEFINITIONS

Locally, the wave Σ can be represented approximately by a plane wave which propagates in the direction $0z$ (Fig. 2.5). The concept of the plane wave allows a series of quantities to be defined which are convenient to use.

The wavelength:

$$\lambda = \frac{v}{v} \tag{2.17}$$

This is the distance travelled in unit time, divided by the number of vibrations in unit time. In free space:

$$\lambda_0 = \frac{c}{v} \tag{2.18}$$

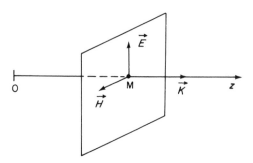

Fig. 2.5. The plane wave. At all points on the plane $z = z_0$, the phases are equal, the amplitudes are possibly equal and propagation is unidimensional

The wave number

$$\frac{v}{v} = \frac{1}{\lambda}$$ (2.19)

This is the reciprocal of the wavelength. As with wavelength, it depends on the chosen unit of length.

The pulsatance, or angular frequency:

$$\omega = 2\pi v$$ (2.20)

In electromagnetism the number 2π appears frequently, and it is preferable to avoid it if possible.

The wave vector $\vec{k}$: along 0z, origin M and modulus or wavenumber

$$k = \frac{\omega}{v} = \frac{2\pi}{\lambda}$$ (2.21)

In free space:

$$k_0^2 = \varepsilon_0 \mu_0 \omega^2$$ (2.22)

In the medium:

$$k^2 = \varepsilon \mu_0 \omega^2 \quad \text{and} \quad k = k_0 \cdot n$$ (2.23 and 2.24)

The phase and group velocities can be expressed by:

$$v = \frac{\omega}{k}$$ (2.25)

and

$$u = \frac{d\omega}{dk}$$ (2.26)

The wave impedance:

$$\frac{E}{H} = \sqrt{\frac{\mu_0}{\varepsilon}}$$ (2.27)

It is expressed in ohms:

$$\sqrt{\frac{\text{Henrys per metre}}{\text{Farads per metre}}} = \text{ohms}$$

The reader will recall that the impedance of a transmission line is:

$$Z = \sqrt{\frac{L}{C}} \text{ohms}$$ (2.28)

From equation (2.27) the intrinsic impedance of free space ($\varepsilon = \varepsilon_0$) is

$$376.73\,\Omega$$

2.8 THE WAVE EQUATION

It can be established from Maxwell's equations that two scalars, E and H relating to the field, are solutions of a second-order partial differential equation, called 'the wave equation' or the Helmholtz equation (Fig. 2.6). For one-dimensional propagation, the case where waves travel in the $0z$ direction, this equation can be written:

$$\frac{\partial^2 \psi}{\partial z^2} = \frac{1}{v^2} \frac{\partial^2 \psi}{\partial t^2} \tag{2.28}$$

All solutions are of the form:

$$\psi(z, t) = f(vt - z) \tag{2.29}$$

in which f is a twice differentiable function of the variable $(vt - z)$. An infinite number of solutions can be written! If two functions $f_1(vt - z)$ and $f_2(vt - z)$ are solutions to equation (2.28), every linear combination $Af_1 + Bf_2$ (where A and B are constants) will also be a solution of equation (2.28).

It can be seen that $\sin[k(vt - z)]$ and $\cos[k(vt - z)]$ are solutions. We can therefore write a sinusoidal function of the form:

$$\psi = A \sin(\omega t - kz) + B \cos(\omega t - kz) \tag{2.30}$$

where $k = \omega/v$, the wave number, and ψ is a wave function, a solution of equation (2.28).

A profile of any wave, like that of Fig. 2.6, can be represented by a series of sinusoidal terms, that is, by a group of waves (theoretically an infinite number) which propagates the profile in the $0z$ direction.

If all the constituent waves have the same velocity v, the profile is conserved, otherwise it disperses. This is the phenomenon of dispersion or dissipation of the group energy, due to the diversity of velocities.

If the profile could be represented by a series of low dissipation constituents, which all propagate at velocities close to the mean v, the profile will be approximately conserved while travelling at the group velocity u of the wave with mean pulsatance ω.

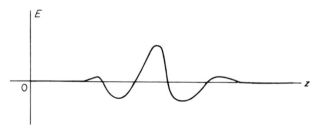

Fig. 2.6. Wave profile at time t

2.9 THE RECIPROCITY THEOREM

Consider a monochromatic electromagnetic (em) source. It is applied to a radiator A_1 and regulated so that A_1 radiates a power W_1. A distant receiver A_2, separated from A_1 by linear media, receives a power W_2. Application of the reciprocity theorem allows the following property to be stated. If, all other things being equal, the em source is applied to A_2 and regulated so that A_2 radiates a power W_1, then A_1 receives a power W_2.

There is reciprocity of propagation between A_1 and A_2 and one can speak of 'attenuation of propagation' between two points or two antennas, without specifying in which direction the propagation occurs. However, certain linear media do not have reciprocity of polarization. When the polarization, that is, the direction of the electric field of the propagating wave, is involved for these media there is reciprocity of the em paths, but there is not reciprocity of attenuation between radiator and polarized receiver.

2.10 GENERAL REMARKS

In electromagnetism it is always dangerous to consider a certain limited domain of space where the field occurs in isolation and to discuss the waves which can be found there. This can only be done if the general and complete solution is already known. To obtain this general solution it is necessary to write Maxwell's equations, then three supplementary linear equations, representing the effect of the material on the field. Therefore there are five equations for five unknown vectors.

The condition for compatibility of these equations allows a dispersion equation to be written, for which the solutions depend essentially on the boundary conditions (physical discontinuities and zero radiation at infinity).

Electromagnetic theory is a powerful means of study, which has given perfectly satisfactory results.

2.11 THE WAVE AND PHOTONS

The electromagnetic wave is associated with photons and it guides them. The intensity of the light is proportional:

(1) To the number of photons, on the one hand; and
(2) To the square of the amplitude of the wave, on the other.

At each point the number of photons is proportional to the amplitude of the Poynting vector, or the square of the amplitude of the electric vector, that is, E^2 (for example), which comes to the same thing. More precisely, the probability that a photon is present at a point, at an instant, is proportional to the square of the amplitude of the wave at that point at that instant.

We have seen that a group of waves, having a small spread around the frequency v, appear as a 'modulation' which propagates with the group velocity

u on the carrier of frequency v. The number of photons will present a volume distribution proportional to the square of the modulated carrier. Thus the packets of photons travel at the group velocity u carried by the 'modulation'.

(3) Light consists of rays

Geometrical optics is based on a number of fundamental principles which have been progressively discovered from direct observation. Here, we shall present several which are particularly useful in the study and use of optical fibres.

The first, most important, most definite and the longest known of these principles can be expressed as follows:

'Light propagates in a straight line in a homogeneous medium.'

That is the principle. It covers all cases if one adds:

'In a homogeneous medium the light from a point source diverges following rectilinear rays.'

There is, in our immediate environment, an object towards which energy naturally converges due to the force of gravity. This object is the Earth. However, this is another matter. We know of numerous sources from which mechanical energy diverges. A pebble falling into still water produces a transverse mechanical wave, a surface wave which diverges. The explosion of a detonator produces a longitudinal mechanical wave which diverges, etc.

If the medium in which the luminous energy propagates is homogeneous, that is, if it has the same properties at all points, one cannot see how the light could propagate other than by the shortest route.

2.12 ATTENUATION AS A FUNCTION OF DISTANCE

Consider a beam of concentric luminous rays from the same point source 0 (Fig. 2.7). Assume that the beam is limited to a solid angle Ω. If Σ_2 is a surface

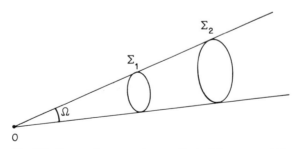

Fig. 2.7. Beam of concentric rays, limited by a cone. Ω is the solid aperture angle of the cone

element within the same cone as surface element Σ_1 with respect to 0, the optical power W_2 which crosses Σ_2 is equal to the optical power W_1 which crosses Σ_1, provided that the propagating medium is neither dissipative, scattering nor luminescent. Hence the energy flux is conserved within the solid angle Ω.

Now assume that the surfaces Σ are spherical and centred at 0. The light flux, per unit area, on Σ_1 will be dW_1/dA_1 and on Σ_2 it will be dW_2/dA_2. It can be seen that the light flux which crosses unit area or 'luminous intensity' is inversely proportional to the square of the distance from the source, r^2:

$$\frac{dW}{dA} \sim \frac{1}{r^2} \tag{2.31}$$

(In electromagnetism, matters are a little less simple at a short distance from the source.)

2.13 REFLECTION AND REFRACTION

In the presence of a physical discontinuity which separates two homogeneous media refraction occurs and can be represented for particular cases such as plane, spherical, parabolic, etc. if the height of surface irregularities is less than $\lambda/100$ (Rayleigh criterion). For a wavelength in air of 1 μm, the maximum height of the irregularity in normal incidence is 100 Å which, for glass, represents a thickness of 30–50 molecules.

Crystals allow much smoother surfaces to be obtained. In oblique incidence an irregular surface appears smoother than in normal incidence, since the apparent height of the irregularity is $h \cos \theta$, where θ is the angle of incidence.

The phenomena of reflection and refraction have been known to man for a very long time, from direct observation. For example, the surface of still water readily produces a horizontal plane mirror in which the image of an object can be seen (a well or sufficiently deep vase with absorbant lining provides a water mirror).

Similarly, for many years man has observed the refraction of light at a junction separating two transparent media such as air and water. For an observer on land a straight stick plunged partly into water appears to be broken exactly at the surface; the wet end appears to be further away from the vertical than the dry end. Similarly, for an observer in the water (for example, a diver looking up) a straight stick, emerging partly into the air, appears broken at the surface; but this time the end in the air appears to be nearer to the vertical than the end in the water. Furthermore, the diver can see only across a limited disc of rays in the air (Fig. 2.8). He observes a transparent region of the air/water surface plane as the inside of a circle centred on the vertical which passes through his eye. Outside the disc the surface of the water appears silvered. In these directions, the diver does not see into the air; the light received by his eyes comes from the water. This very important phenomenon is called 'total internal reflection'.

Descartes' laws on reflection and refraction concern the interface plane which

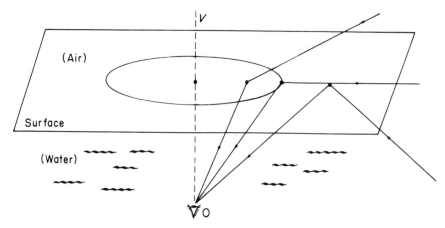

Fig. 2.8. What does an observer immersed in water see in the air? The observer at 0 looks upwards. In the air he can see only across a circular disc of rays limited by total internal reflection. Outside, the surface of the water appears to him to be metallized and, in these directions, he receives only light which originates in the water

separates two homogeneous and isotropic media. Consider a ray in medium (1) incident on the interface at M. This ray and the normal through M define a plane of incidence. It results in a reflected and/or refracted ray at point M. These two rays, when they exist, are in the plane of incidence. Furthermore, the incident ray and the reflected ray, when it exists, make equal angles with the normal. Finally, the ratio of the sine of the angle of incidence and the sine of the angle of refraction, when it exists, is constant:

$$\theta_1 = \theta'_1 \qquad (2.32a)$$

$$\frac{\sin \theta_1}{\sin \theta_2} = \text{Constant} \qquad (2.32b)$$

This constant is called the relative refractive index of medium (2) with respect

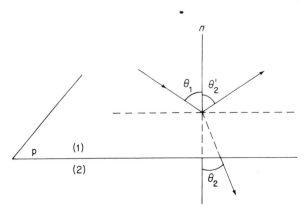

Fig. 2.9. Rays reflected and refracted at the (1–2) interface .

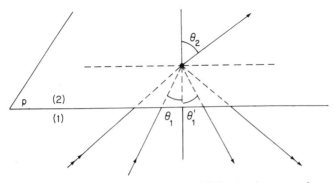

Fig. 2.10. Reflection and refraction of light. In the case where $n_1 > n_2$, a limiting angle of incidence θ_c exists above which $(\theta_1 \geqslant \theta_c)$ there is no refracted ray

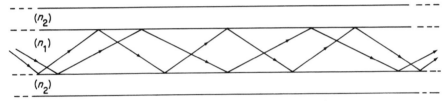

Fig. 2.11. Light guiding in a ribbon. A layer of refractive index n_1 is sandwiched between two layers of refractive index $n_2 < n_1$. The totally internally reflected rays propagate without loss on reflection

to medium (1). The refractive index of a medium relative to free space is called the absolute refractive index. The relative index of two media is the ratio n_2/n_1 of the absolute indices (Figs 2.9–2.11).

It is clear that equation (2.32b) can be satisfied only for a limited range of values of θ_1 if the constant is less than unity. The refracted ray does not always exist. In the case where it does not, the reflection is described as 'total'.

2.14 TOTAL INTERNAL REFLECTION

The phenomenon of total internal reflection is most important for the guiding of light. It allows propagation of rays by total internal reflection, in a broken line, in a layer of material of high index n_1 between two layers of material of low index n_2.

The critical angle at which reflection becomes total is given by:

$$\theta_c = \arcsin \frac{n_2}{n_1} \tag{2.33}$$

For $\theta_1 \geqslant \theta_c$, total internal reflection occurs.

Assuming that medium (2) is air of unity refractive index, the values of θ_c are indicated in Table 2.1 for several transparent materials.

Table 2.1

Material	Critical angle θ_c (degrees)
Water	48.6
Ordinary glass	41.8
Crystal glass	31.8
Diamond	24.4

It can be seen in the table that diamond is a real light trap. From the inside of the stone only rays with a direction near to the normal to one facet can escape. Those which have an incidence greater than 24.4 degrees are totally internally reflected and the light stays within the diamond. The optical power density—in other words, the number of photons per unit volume—is much greater inside the diamond than in the external air. If, on the other hand, the cut of the stone favours light output by one or several privileged facets—these sparkle. The same applies to cut crystal glass but to a lesser extent.

2.15 THE PRINCIPLE OF CONSERVATION OF EXTENT

The other fundamental principles of geometrical optics are Fermat's principle, or the stationary phase principle, Huygen's construction and Malus's theorem. It can be shown that these principles are equivalent to each other and to Descartes's law. They will not be developed here; however, one last principle, that of the conservation of extent, will be quoted.

Consider a real object of area A_1 emitting light in air within the solid angle Ω_1 towards an optical system. After a series of several reflections and refractions the optical system produces an image of area A_2 within the solid angle Ω_2 such that:

$$A_2 \cdot \Omega_2 = A_1 \cdot \Omega_1 \tag{2.34}$$

Whatever the intermediate system may be, the product $A \cdot \Omega$ cannot be increased

(4) Extension of scalar optics

2.16 FRESNEL'S REFLECTION COEFFICIENTS

Descartes's laws are fortunately complemented by Fresnel's formulae concerning reflection coefficients, which are part of electromagnetism, and they will not be developed here. We shall indicate only that the reflection coefficient of optical power for normal incidence on a plane surface is:

$$R = \left(\frac{n_1 - n_2}{n_1 + n_2}\right)^2 \tag{2.35}$$

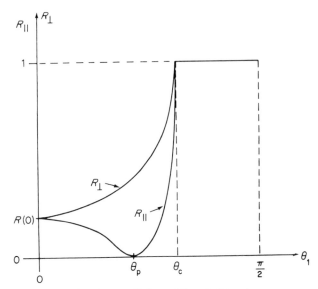

Fig. 2.12. Reflection coefficients $R\parallel$ and $R\perp$, when $n_1 > n_2$. The Brewster angle is reached for $\theta_1 + \theta_2 = \pi/2$, before the incidence of total internal reflection $\theta_2 = \pi/2$ as θ_1 increases

It is to be noticed that permutation of n_1 and n_2 does not change the result. Hence, on a lamina with plane and parallel faces the coefficient is the same for interface (1–2) and interface (2–1).

For a window of ordinary glass one obtains 4% for each face and 7.7% for both faces. The rest of the light, that is, 92.3%, in normal incidence is transmitted through the glass (Fig. 2.12.). However, we shall add several useful results. With the same notation as above and for $n_1 > n_2$, total internal reflection occurs for $\theta_1 \geqslant \theta_c = \arcsin(n_2/n_1)$. Hence, the Fresnel reflection coefficient of optical power is equal to 1 throughout the interval $\theta_c \leqslant \theta \leqslant \pi/2$. However, when reflection is not total, it is necessary to distinguish between two cases:

1st case: The electric field E of the light wave is parallel to the plane of incidence. In this case, the coefficient is denoted by $R_\parallel$.

2nd case: The electric field of the wave is perpendicular to the plane of incidence. The coefficient is denoted by $R_\perp$.

It can be shown that $R_\perp$ does not become zero for any value of θ_1.

Table 2.2

Material	Index n_1	R(0)	θ_p (degrees)	θ_c (degrees)
Water	1.333	0.02	37	48.6
Ordinary glass	1.50	0.04	33.7	41.8
Crystal glass (flint)	1.90	0.10	27.8	31.8
Diamond	2.417	0.17	22.5	24.4

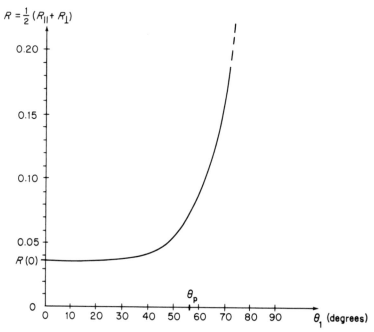

Fig. 2.13. Coefficient of reflection of energy from a non-polarized source by a plane interface separating air of refractive index 1 and glass of refractive index $n_2 = 1.48$. The remarkable constancy of the coefficient is evident; it remains between $R(0) = 0.0374$ and $R(35) = 0.0404$ for 35 degrees variation of θ_1. From $\theta_1 = 0$ to $\theta_1 = 25$ degrees the coefficient changes from 0.0374 to 0.0381. The Brewster angle is near to 56 degrees. *Conclusion*: For non-polarized light, the loss on reflection at a plane air/glass interface is small and remarkably constant near to normal incidence

It contrast, the coefficient $R_{\parallel}$, given by Fresnel's equation, becomes zero for $\theta_1 + \theta_2 = \pi/2$; this equation always has a solution $\theta_1 = \theta_p$, in the interval $0 < \theta_1 < \pi/2$. This incidence angle θ_p is called the Brewster angle.

Assuming that medium (2) is air of refractive index unity, Table 2.2 gives values of the reflection coefficient $R(0)$ for normal incidence, the Brewster angle θ_p and the critical angle θ_c for different materials (Fig. 2.13).

2.17 DIFFRACTION BY APERTURES

If a beam of light illuminates an obstacle in which there is a small aperture, a diverging pencil of rays, which have penetrated the aperture, is observed behind the obstacle (Fig. 2.14). In other words, luminous energy is found in the geometric shadow of the obstacle. This phenomenon is called 'diffraction'. It is very general and is a consequence of the wave nature of light. It is observed in acoustic and other wave propagations (see note at the end of this chapter).

In coherent light an expression for the angular divergence of the pencil

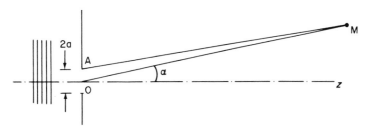

Fig. 2.14. A coherent source of light, represented as a series of planes of
equal phase, in normal incidence on an obstacle pierced with a circular
aperture of radius a. It is diffracted by the aperture

transmitted through a circular aperture is obtained by observing that the field
at M can have a significant intensity only if the ray from side A of the aperture
is approximately in phase with the ray from the centre 0. Hence

$$\alpha \simeq \lambda/a \qquad (2.36)$$

Three regions can be distinguished behind the obstacle:

(1) The near field;
(2) The intermediate or Fresnel zone; and
(3) The distant or Fraunhofer zone.

The diameter of the diffracted pencil becomes significantly different from that
of the aperture only at a distance d which is of the order of a^2/λ. For example, for
$\lambda = 1\ \mu$m and $a = 1$ mm, $d = 1$ m. The centre of the intermediate region is situated
at this distance of 1 m.

At a great distance, the wavefront is approximately spherical and centred at
0. The beam thus resembles a pencil of concentric rays. Notice that the expression
$\alpha \simeq \lambda/a$ is applicable only if a is much greater than λ.

Babinet's principle: It is possible to determine, by calculation, the true shadow
cast by an object by assuming that the obstacle is an aperture and that this
aperture radiates darkness. It is found, for example, that a human hair
illuminated with red light around 700 Å casts a shadow of which the
aperture is around 1 degree.

(*Note*: Vibration of the superficial atoms and molecules, caused by the light,
can extend to the other face of the obstacle and cause an edge effect.)

3 The step index (SI) fibre

3.1 THE GUIDING STRUCTURE

The SI fibre is the optical fibre in current use, and the light guide is fabricated in the form of a cylindrical interface of radius a (Fig. 3.1). The included light propagates with little loss in this cylinder, which is called the *core*; it consists of an amorphous, transparent and homogeneous material of refractive index n_1. The core is surrounded by a *cladding* of amorphous, transparent and homogeneous material of refractive index n_2 which forms an optical interface permitting total internal reflection ($n_1 > n_2$). Some rays, called 'guided' rays, propagate in the core along a broken line consisting of successive total internal reflections along the cylindrical interface.

This guidance can be explained by an analogy. Consider a vehicle track consisting of three bands on which small three-wheeled cars circulate. The two wheels in front are driven and their speed of rotation depends on the band on which the tyre finds itself; this is greater on the outside bands than on the central one. When a front wheel of a car running on the central band encounters an outside one this wheel will accelerate and the car will find itself returned to

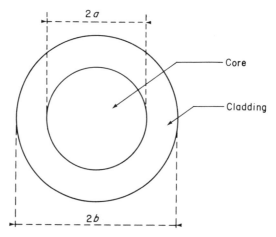

Fig. 3.1. Diagram of the cross-section of a SI multi-mode fibre. The diameter of the core is 2 a

34

the central band. For a range of departure directions, the cars will be guided along a zig-zag path between the side bands.

3.2 MERIDIONAL CORE RAYS

We shall consider a rectilinear fibre; more precisely, a cylindrical rod along the $0z$ axis consisting of a core of radius a and an exterior cladding of radius b. In this cylindrical model we shall consider meridional planes passing through $0z$ and the light rays in these planes. With the exception of rays parallel to $0z$, they all intercept this axis. The important parameter is the angle θ which each of these rays makes with $0z$ (see Fig. 3.2).

Consider a meridional ray which makes an angle θ with $0z$. The angle ϕ which this ray makes with the normal to the interface, assumed to be plane at the point of intersection B, is $\phi = \pi/2 - \theta$. If ϕ is sufficiently small, that is, θ is sufficiently large, there will be a refracted ray in the cladding in addition to a reflected ray R_1 in the core. Energy is therefore transmitted across the interface and can be lost. In contrast, if $\phi \geqslant \phi_c = \arcsin n_2/n_1$, there is no refracted ray R_2 and no power loss; all the power is reflected and conserved within the core.

It is clear that ray R_1 will, in turn, be totally internally reflected by the interface and propagation will continue in a broken line, formed of equal segments inclined at an angle θ to $0z$, theoretically without loss.

The guidance condition for meridional rays in the core is therefore:

$$\theta \leqslant \theta_c = \arccos \frac{n_2}{n_1} \qquad (3.1)$$

All meridional rays, which are incident at any point on the circular disc of the core (radius a and centred on $0z$) of an input section of the cylindrical rod, will be guided if their corresponding angle of refraction is less than θ_c; that is, if the angle of incidence θ_0 in the exterior medium of refractive index n_0 is less than a certain limiting angle of acceptance θ_{ACC}.

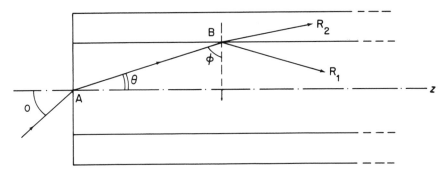

Fig. 3.2. Reflection of a ray of light on the cylindrical core–cladding interface. The incident ray at an angle ϕ to the interface, assumed to be locally plane, gives rise to a reflected ray R_1 and also, in the general case, a refracted ray R_2 in the cladding in according with Descartes's law for the plane interface

The reader will recall that:

$$n_1 \sin \phi_c = n_2 \quad n_1 > n_2 \tag{3.2}$$

$$n_1 \cos \theta_c = n_2 \tag{3.3}$$

$$n_0 \sin \theta_{\text{ACC}} = n_1 \sin \theta_c = NA \tag{3.4}$$

The quantity $n_1 \sin \theta_c$ from Descartes's law (which is independent of n_0) is called the 'numerical aperture' (NA). It is found that:

$$NA = n_1 \sqrt{\left(\frac{n_1^2 - n_2^2}{n_1^2}\right)} \tag{3.5}$$

Normally one puts:

$$2\Delta = \frac{n_1^2 - n_2^2}{n_1^2} \tag{3.6}$$

and writes:

$$NA = n_1 \sqrt{(2\Delta)} \tag{3.7}$$

Finally, the *guidance condition* can be written:

$$\sin \theta \leqslant \sqrt{(2\Delta)} \tag{3.8}$$

(*Note 1:* For small values of Δ, that is, for a small relative difference of refractive index between core and cladding, one has, approximately:

$$\Delta \simeq \frac{n_1 - n_2}{n_1} \tag{3.9}$$

Note 2: The angles ϕ_c and θ_c are often called 'critical' angles.)

3.3 HELICAL CORE RAYS

Consider again the cylindrical rod along the $0z$ axis with core radius a. The refractive index of the core is n_1 and the index of the cladding is n_2. Rays which are in the core and do not meet the $0z$ axis are called helical or 'skew' rays. They are much more numerous than meridional rays.

If we consider a point P, off the $0z$ axis of the cylindrical rod of radius a, we can characterize an optical ray R passing through P by the angle θ which it forms with the $0z$ axis and by the shortest distance a_0 between this ray and the $0z$ axis (Fig. 3.3). For meridional rays the distance a_0 is zero, for helical rays it is not.

Let us project these elements on to the cross-sectional plane of the rod through 0. The interface projects as a circle of centre 0 and radius a. The point P projects as p. The segment of the optical ray passing through P becomes a chord of a circle of radius a passing through p and tangential to a circle of radius a_0 ($a_0 \leqslant 0p$).

If one considers only helical rays propagating in the direction of increasing

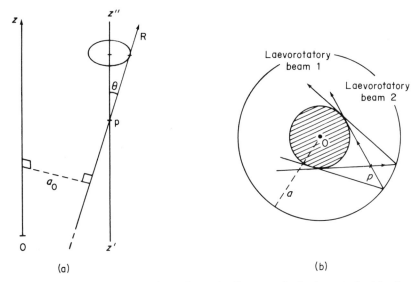

Fig. 3.3. (a) A ray R, passing through a point P not on $0z$, is characterized by the angle θ which it makes with $0z$ and by its minimum distance a_0 from $0z$. (b) On the plane of cross-section of the rod passing through 0, the point P is projected on to p. A rectilinear segment of the ray passing through P is projected following a chord of the circle of radius a, passing through p and tangential to the circle of radius a_0

z it can be seen that for a given value of a_0 there are four helical rays, two laevorotatory and two dextrorotatory, passing through p. For $a_0 = 0p$ there are only two distinct helical rays, one laevorotatory and one dextrorotatory.

More precisely, consider three consecutive reflections at successive points L, M and N on the cylindrical interface (Fig. 3.4). Project the rays on to the cross-sectional plane of the cylinder through M with centre 0. The projections lM of LM and Mn of MN each form an angle φ with 0M and are, furthermore, tangential to the circle of radius a_0. It can be seen that the projection of a helical ray is formed of equal chords, each tangential to the circle of radius a_0 and each of which is derived from its predecessor by a rotation of $(\pi - 2\varphi)$ about 0. It is clear that the angle φ is conserved at each reflection and from one reflection to the next.

It can be shown that the guidance condition for total internal reflection of a helical ray is:

$$\sin \theta \cos \varphi \leqslant \sqrt{(2\Delta)} \qquad (3.10)$$

In conclusion, the guidance condition $\theta \leqslant \theta_c$, which is strictly for meridional rays and which is written:

$$\sin \theta \leqslant \sqrt{(2\Delta)} \qquad (3.8)$$

is superfluous for helical rays. However, it allows a simple calculation of the

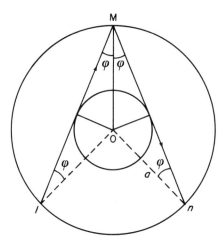

Fig. 3.4. Consider three consecutive reflections at the points L, M and N on the interface $r = a$. If the projections lM of LM and Mn of MN are formed on the cross-section through M, each forms an angle φ with OM and is tangential to the circle of radius a_0. A helical ray is formed with segments of equal length each making an angle θ with $0z$ and turning, on each reflection, through $(\pi - 2\varphi)$ about $0z$

acceptance angle and an estimate of the optical power propagated in the core of a step index fibre.

3.4 CLADDING RAYS

Propagation in partial reflection, with rays refracted in the cladding, occurs when the incidence angle of the ray on the interface cylinder is less than ϕ_c. In this case, part of the energy penetrates into the cladding. For propagation to a great distance this energy is generally lost. However, it can be reflected from the interface $r = b$, at least partially, to return to the core and enter it again in a repetitive core-cladding path strewn with partial reflections.

At a short distance from the input section cladding rays often participate greatly and play a major part in laboratory experiments (Fig. 3.5).

In the presence of an opaque packaging surrounding the cladding all cladding rays are leakage rays. Only core rays are to be considered at large distances. It is for them that a core–cladding interface has been made with a cladding of a quality which does not attenuate the totally reflected ray. Electromagnetism shows us that if the light propagates in the core there must be some in the cladding (Fig. 3.6).

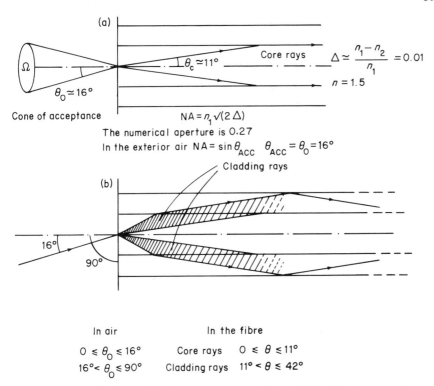

Fig. 3.5. Numerical example relating to a current fibre. (a) Meridional rays guided
by the core on which the transmission of signals depends; (b) cladding rays. These
supplementary rays make an important contribution over short distances

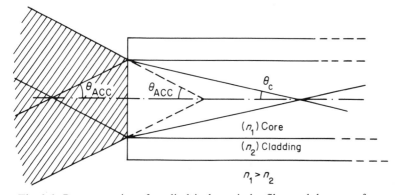

Fig. 3.6. Representation of a cylindrical step index fibre and the cone of rays
accepted by the core and which fill it with light propagated by total internal
reflection, practically without loss, if the cladding is not too thin

3.5 THE EXISTENCE OF PROPAGATION MODES

Physics teaches us that all physical structures can be quantified. Propagation of light in a cylindrical dielectric rod which represents an optical fibre can therefore be quantified. It is expressed as a dispersion equation of which all solutions are not permissible.

The analysis introduces wave functions, each of which is one of the solutions of the problem and is characterized by two integers which define its structure. These functions are called 'modes'. In a step index fibre there can be very many and they can be superposed. The number of them, which is an integer, is approximately:

$$N \simeq \tfrac{1}{2}F^2 \tag{3.11}$$

If one takes account of laevorotatory and dextrorotatory polarization, this number is doubled.

F is called 'the normalized frequency'. It is a parameter which, on its own, defines the propagation conditions of the light in the rod:

$$F^2 = a^2(k_1^2 - k_2^2) \tag{3.12}$$

where $2a$ is the diameter of the rod, k_1 the wave number in the medium which constitutes the rod and k_2 the wave number in the exterior medium.

A result of this form can be compared with those of geometrical optics. To make the comparison, we write N in the form

$$N \simeq \frac{1}{2}\left(\frac{2\pi a}{\lambda_0}n_1\right)^2 \cdot 2\Delta \tag{3.13}$$

In geometrical optics, propagation is independent of a, b and λ_0 (to be rigorous, $\lambda_0 = 0$ is assumed). Guidance is limited to rays forming an angle $\theta < \sqrt{(2\Delta)}$ (for small values of $\sqrt{(2\Delta)}$ with the $0z$ axis.

In electromagnetic theory there is a finite number of guided modes. In fact, the difference between the results obtained by the two methods is of practical significance only if the number of guided modes is small (less than several tens).

3.6 NUMERICAL APERTURE

In conclusion, a multimode SI fibre is characterized primarily by its numerical aperture:

$$NA = n_1\sqrt{(2\Delta)} \qquad 2\Delta = \frac{n_1^2 - n_2^2}{n_1^2} \tag{3.6 and 3.7}$$

The angle of acceptance is given by:

$$n_0 \sin\theta_{ACC} = NA \tag{3.4}$$

where n_0 is the index of the exterior medium. The solid angle formed by the cone of acceptance is:

$$\Omega_{ACC} = 2\pi(1 - \cos\theta_{ACC}) \qquad (3.14)$$

3.7 ATTENUATION AS A FUNCTION OF DISTANCE AND BENDING LOSS

The different propagation modes in the fibre are not attenuated in the same proportion per unit distance. However, there is a mixture of energies propagated by the modes caused by the inevitable imperfections in the production of the guide. The result of this mixture is the formation, at a distance from the input which varies from hundreds of metres to one or two kilometres, of an 'equilibrium train of modes' for which the attenuation per unit distance is constant. One of the important consequences of this mixing of modes is the suppression of path reciprocity. Reciprocity certainly exists for each individual mode but their interaction does not allow practical use. In conclusion, a fabricated optical fibre is characterized by a linear attenuation A of optical power, in decibels per kilometre.

3.7.1 Bending loss

So far we have considered propagation of light in a rectilinear cylinder. In fact the fibre can conveniently be curved and the loss in following the curvature is very small.

In the curved case (Fig. 3.7) the core rays, which are close to the critical angle ϕ_c, are partially refracted and lose some of the guided energy. Along the length

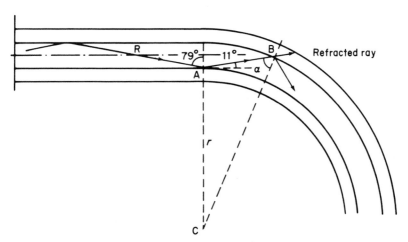

Fig. 3.7. Bending loss. In the rectilinear part of the fibre, the ray R guided by the core is reflected at A with an incidence of 79 degrees. In the curvilinear part of the fibre the ray reflected at A is incident at B with an angle α less than 79 degrees. There is therefore a refracted ray in the cladding and a loss

of the curved trajectory (for example, in an optical cable where the fibre is twisted into a helix) the linear attenuation α is increased by $\Delta\alpha$. The mean quantity $\Delta\alpha$ (dB/km) can be calculated from the approximate relation:

$$\Delta\alpha = 10\log\left[1 - \frac{2an_1^2}{(NA)^2 . r}\right] \tag{3.15}$$

where r is the radius of curvature.

In practice, for $r \geqslant 1$ cm the bending loss is very small. (For $2a = 75\,\mu$m, the ratio $2a/r$ is less than or equal to 75.10^{-4}. The loss depends on the ratio expressed by

$$2\Delta = \frac{n_1^2 - n_2^2}{n_1^2} \geqslant 10^{-3}$$

3.8 THE DATA RATE CAPACITY

The data rate of any fibre is limited by the phenomenon of dispersion. (For the moment we shall not consider rate limitations imposed by the light source and detector but only the limitations imposed by a linear, multimode and dispersive light guide.)

If a hypothetical impulse, of negligible duration and consisting of a group of light waves having a small spread about the frequency v, enters the fibre by filling the cone of acceptance with the same intensity in each direction the different rays propagated will produce an impulse at the output which is spread with respect to time. The luminous energy will be dispersed between the shortest and longest trajectories according to the group velocity $u(v)$ (Fig. 3.8).

The group time on the shortest trajectory is:

$$t = \frac{L}{u(v)}$$

where L is the length of the fibre. The group time on the longest trajectory is:

$$t_c = \frac{L}{\cos\theta_c}\frac{1}{u(v)}$$

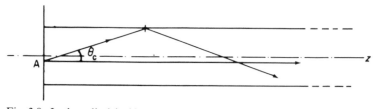

Fig. 3.8. In the cylindrical homogeneous rod the ray parallel to the axis has the shortest trajectory and the critical ray, which forms an angle θ_c with the axis, has the longest one

The time spread between the groups is then:

$$\Delta t = t_c - t = \frac{L}{u(v)}\left(\frac{1}{\cos\theta_c} - 1\right) \qquad (3.16)$$

and the spread τ per unit length, due to modal dispersion, can be written:

$$\tau = \frac{\Delta t}{L} = \frac{1}{u(v)}\left(\frac{n_1}{n_2} - 1\right) \qquad (3.17)$$

If, modifying the previous assumptions, it is assumed that the light impulse at the input is still of negligible duration but of a larger spectral spread from v_1 to v_2, the group time on the shortest trajectory, covered at the highest velocity, will be:

$$t = \frac{L}{u(v_1)}$$

if

$$u(v_1) > u(v_2)$$

(It is assumed that the function $u(v)$ is monotonic and decreasing in the interval v_1, v_2.)

The spread per unit length τ will be given by:

$$\tau = \frac{\Delta t}{L} = \frac{1}{u(v_2)}\cdot\frac{n_1}{n_2} - \frac{1}{u(v_1)} \qquad (3.18)$$

This is the expression for the spread in duration, per unit length, of the impulse received at the output of a multimode SI fibre, caused by modal and material dispersion, when the light source is not monochromatic.

However, if the material which constitutes the fibre can be considered to be non-dispersive in the interval (v_1, v_2), the material dispersion can be neglected, and one puts:

$$u(v_1) = u(v_2) = v = \frac{c}{n_1}$$

From which:

$$\tau = \frac{1}{c}n_1\sqrt{(2\Delta)} \qquad (3.19)$$

In practice, silica has zero dispersion in the vicinity of $1.25\,\mu m$ for the wavelength λ_0 and, with a suitable doping one can obtain minimal material dispersion at any required wavelength between 1 and $1.6\,\mu m$.

The principal cause of spreading of impulses in SI fibres is modal dispersion. For short distances, equations (3.18) and (3.19) adequately indicate the spread $\Delta t = L\cdot\tau$ up to at least several hundred metres and, at most, several kilometres. At greater distances, dissipation and intermodal coupling greatly modify the form of the received impulse. It forms an equilibrium train of modes which is revealed in experiments which real fibres.

44

3.9 MODES PROPAGATED IN REAL FIBRES

Coupling of modes is caused by curvature of the cable, microcurvature of the fibre in the cable, imperfections of the core–cladding interface (variation of diameter and ellipticity) and, more generally, scattering due to irregularities of the material.

A guided mode of any rank can give up energy to modes of lower rank and to modes of higher rank, which are leakage modes. Modes of higher rank disappear and modes of lower rank are attenuated. The optical energy is then distributed among an 'equilibrium train of modes'.

Figure 3.9 shows the effect of variations of the diameter 2a of the core, which can be 1 μm, for example. Variation of diameter, as with microcurvature, causes

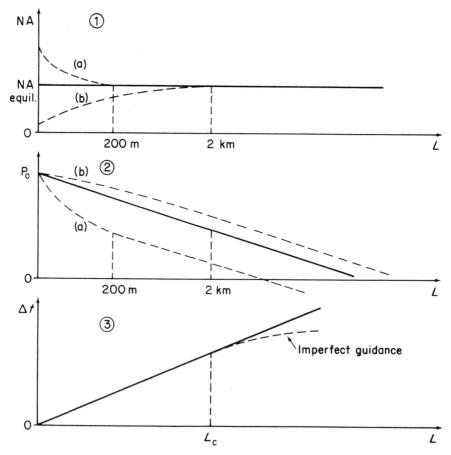

Fig. 3.9. Propagation in real, imperfect, fibres. Variation as a function of distance L of the numerical aperture (1) and optical power (2) for an excitation which is (a) very divergent and (b) very directional. (3) is the variation of pulse spreading as a function of L with (broken line) and without coupling

a loss of light. In effect the cone of guided rays is effectively limited to a certain angle θ_{eff}, less than the theoretical angle θ_c.

Also, irregularities of the core–cladding interface often form repetitive structures as a function of distance. They play the part of a diffraction grating which has preferential modes and causes coupling between the modes. The overall energy distribution tends towards an equilibrium state.

Figure 3.9(1) shows the effect of variation of the aperture angle of the cone of guided rays as a function of distance when the injection of light is excessive (that is, too divergent) and, in contrast, insufficient.

Figure 3.9(2) shows variations in the optical power P_0 as a function of distance and also when injection is excessively or insufficiently divergent.

For the two cases of injection the same equilibrium of modes is reached which presents an equilibrium numerical aperture NA_{equil} and an equilibrium linear attenuation α_{equil}.

If light is injected into an SI fibre within the effective solid acceptance angle at equilibrium Ω_{equil}, the attenuation is α_{equil} from the start and the equilibrium numerical aperture applies throughout the fibre. (This assumes that fabrication of the fibre and the cable is independent of time and fault-free.)

3.10 THE PRACTICAL CONSEQUENCES OF COUPLING OF MODES

3.10.1 Over short distances ($L < L_c$)

It is advisable to use theoretical formulae to estimate performance. Capacity is calculated using equation (3.19) and then (3.18), which will provide the spread of the received pulsewidth $\Delta t = \tau \cdot L$, generally expressed in nanoseconds per kilometre (ns/km).

The optical power entering the fibre by the cone of acceptance can be estimated by calculation or measured on a short segment (Fig. 3.9).

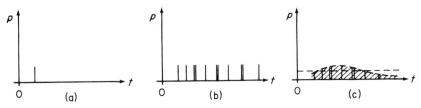

Fig. 3.10. Optical pulse received at the output of a guide. The input pulse is of quasi-zero duration and very narrow spectral bandwidth. (a) Monomode guide—the output pulse is of quasi-zero duration and unique; (b) multimode guide without dissipation and without coupling—the output pulse is theoretically composed of a limited series of pulses of quasi-zero duration; (c) a real multimode guide—dissipation and intermodal coupling mix the modes. The optical power detected is in the form of an elongated current pulse

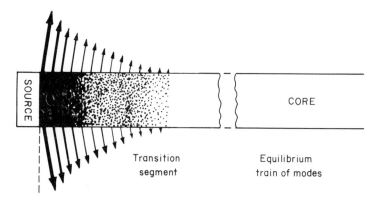

Fig. 3.11. The source injects light into the core of the fibre and stimulates leakage modes which disappear after a certain distance

3.10.2 Over long distances ($L \geqslant L_c$)

In the equilibrium train of modes photons pass from slow to fast modes and vice versa (Fig. 3.10); this reduces the distance between the first and last arrivals. If the length L of the fibre is greater than the coupling length L_c, the spread of the received pulse is given by:

$$\Delta t = \tau \sqrt{(L_c \cdot L)} \, (\mathrm{ns/km}) \qquad (3.20)$$

Thus the intermodal dispersion ceases to increase linearly when L exceeds L_c.

Unfortunately, manufacturers do not always specify the coupling length L_c, because this characteristic is rather variable between measurement in the laboratory and measurement on an installed cable. For very high-quality fibres L_c is greater than 1 km and can reach 10 km if the fibre has very little loss ($\alpha \leqslant 0.6 \, \mathrm{dB/km}$) (Fig. 3.11).

For ordinary fibres L_c is smaller. For example:

NA	NA (equil.)	θ (equil.) degrees	L_c(m)
0.30	0.08	3	530
0.16	0.09	3.5	130

It should be noticed that coupling of modes is a favourable phenomenon when the optical link is limited in distance by an adverse required data rate and not by the optical power necessary for reception, since the coupling of modes loses energy and increases the linear attenuation coefficient α.

4 Propagation of light in a homogeneous dielectric rod; the classical point of view

4.1 INTRODUCTION

This chapter gives the reader an overview of our knowledge of propagation of light in a step index fibre based on classical electromagnetic theory. Attention will be restricted to a pure non-dissipative dielectric rod within an unlimited non-dissipative dielectric medium. Consequently, in the two media, the current density is zero ($\sigma \equiv 0$) and the electrical permittivity ε is real. If guidance occurs there is no attenuation. There is neither absorption or scattering.

Regardless of the terms used to describe it, the rod represents the core of a fibre and the external medium represents the cladding. A complete, exact solution exists for the step index model:

$$n(r) = \begin{cases} n_1 & r \leqslant a \,(\text{core}) \\ n_2 & r > a \,(\text{cladding}) \end{cases} n_1 > n_2 \qquad (4.1)$$

where n is the refractive index and r is the distance from the axis of revolution Oz of the rod of diameter $2a$. Both the rod and the exterior medium are homogeneous.

Starting with Maxwell's equations or the Helmholtz wave equation:

$$\Delta \vec{\psi} - \varepsilon\mu_0 \frac{\partial^2 \vec{\psi}}{\partial t^2} = 0 \qquad (4.2)$$

where $\vec{\psi}$ indicates a component of the electric or magnetic field and with boundary conditions of $r = a$ at the interface and zero radiation for $r = \infty$, a harmonic dispersion equation of angular frequency ω is obtained.

The solutions of this equation are Bessel functions of the first kind in the core J_ν and of the second kind modified in the cladding K_ν. Each of these pairs of solutions constitutes a group of modes which, for the core, can be decomposed into a finite series of possibilities. Hence each mode is characterized by two integer parameters ν and μ.

47

4.2 THE AXIAL WAVE NUMBER β

Even if they have the same colour or frequency, the velocity of photons along $0z$ depends on the mode which guides them. This fact is expressed by the phase factor:

$$e^{j(\omega t - \beta_m z)}.$$

in the expression for the field of angular frequency ω, relative to the mode m. β_m denotes the wave number of mode m along $0z$:

$$\beta_m = \left(\frac{\omega}{v_z}\right)_m.$$

For guided modes, β_m is a real number which cannot take an arbitrary value. Permitted values of β_m depend on the integer parameters v in azimuth and μ in radius.

Furthermore, the phase velocity v_z can be neither greater than v_2 nor less than v_1, the wave number $k_2 = \beta_m$ must satisfy the double inequality:

$$k_2 < \beta_m < k_1 \qquad (4.3)$$

The series of solutions is therefore limited.

Each solution can be computed as a function of a single unique variable which contains the data of the problem. These are:

(1) The diameter $2a$ of the core; and
(2) The wave numbers $k_1 = k_0 \cdot n_1$ and $k_2 = k_0 \cdot n_2$ of the media of the core and cladding, respectively.

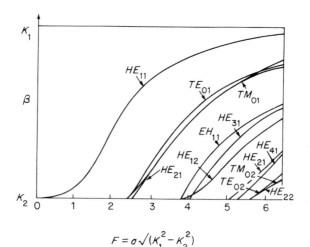

$$F = a\sqrt{(K_1^2 - K_2^2)}$$

Fig. 4.1. Variation of axial wave number β of the first modes of propagation as a function of the normalized frequency F in a step index rod. (Exact solution)

This variable F is called the 'normalized frequency' and is defined by:

$$F = a(k_1{}^2 - k_2{}^2)^{1/2}. \tag{4.4}$$

For two propagating media of given refractive indices n_1 and n_2, the relevant variable is a/λ_0:

$$F = \frac{a}{\lambda_0} 2\pi \cdot n_1 \sqrt{(2\Delta)} \tag{4.5}$$

4.3 GRAPHICAL REPRESENTATION OF EXACT SOLUTIONS

Figure 4.1, limited to $F \leqslant 6.5$, shows all the solutions to the problem graphically (that is, all the guided modes). It clearly illustrates that guiding of modes can exist only for values of F^* greater than a certain limit F_c. This phenomenon is called 'cutoff'. The curve is cut by the line $\beta = k_2$. Below the cutoff, the mode escapes radially, so divergence is not totally suppressed. In consequence of the phenomenon of cutoff, guided modes become less and less numerous as F decreases.

The fundamental mode is that for which $F_c = 0$, and it exists for all values of F. It remains the only guided mode in the interval:

$$0 < F < 2.405 \tag{4.6}$$

This mode is a necessary result of the theory of guidance.

For this mode an approximate solution exists which allows β to be estimated as a function of F (due to Gloge):

$$u = \frac{(1 + \sqrt{2})F}{1 + (4 + F^4)^{1/4}} \tag{4.7}$$

and

$$\beta^2 = k_1^2 - \frac{u^2}{a^2}$$

Hence, as F increases from zero, more and more curves appear on the line $\beta = k_2$. They are continuously increasing with F and tending towards a common asymptote $\beta = k_1$.

These modes are denoted by two letters and two integer index numbers. The first of these numbers is the azimuth parameter v. The azimuthal period is an integral fraction of 2π: $2\pi/1$, $2\pi/2$, $2\pi/3$,...

The second number is the radial parameter μ, which, given the oscillatory nature of Bessel functions of the first kind, defines the series of solutions u_{v1}, u_{v2}, u_{v3},..., etc., which give the same value to $Jv(u \cdot r/a)$.

When a longitudinal component of the field E_z or H_z is zero, the mode is said to be 'transverse electric' (TE) or 'transverse magnetic' (TM). When E_z and H_z both exist, the mode is said to be 'hybrid'. Modes are called EH or HE

---

* The normalized frequency is denoted by V by many authors.

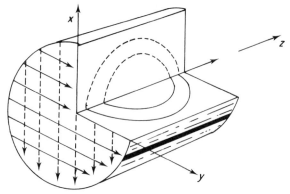

Fig. 4.2. States of polarization of the fields for the HE_{11} mode of a cylindrical rod. (E field in full line, H field in broken line)

according to which of E_z and H_z is dominant. The fundamental mode is the hybrid $HE_{11}(H, E, \text{one}, \text{one})$. It exists with left- and right-handed circular polarizations, as all the modes (Figs 4.2 and 4.3).

4.4 THE NUMBER OF GUIDED MODES

The number of guided modes is a finite integer. It can be shown that it is approximately equal to:

$$N = \tfrac{1}{2}F^2 \qquad (4.8)$$

$F >$	0	2.405	3.832	5.136	5.520	6.380	7.016
$N \geqslant$	1	3	6	8	10	12	15

For multimode fibres, the number of guided modes can reach several thousand

4.5 MODAL DISPERSION

Figure 4.1 again allows the series of values of β_m at a given high frequency to be represented. The highest value of β is that of the fundamental mode; this is followed by the modes of low rank, at first very close to each other, and finally, the modes of high rank with increasingly greater separation.

An intermodal dispersion of phase velocity $(v_z)_m$ therefore exists:

$$\frac{1}{(v_z)_m} = \frac{\beta_m}{\omega} \qquad (4.9)$$

For a luminous ray having a small spread in the vicinity of λ_0, that is, for

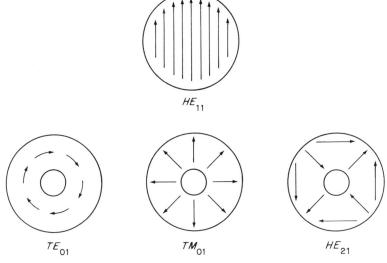

Fig. 4.3. Polarization states of the electric field for the first guided modes of
a cylindrical rod

a normalized frequency having a small spread about F, a dispersion of group
velocities $(u_z)_m$ also exists:

$$\frac{1}{(u_z)_m} = \frac{d\beta}{d\omega} = \frac{d\beta}{dF} \cdot \frac{dF}{d\omega} \tag{4.10}$$

where $d\beta/dF = C_m(F)$ is a characteristic of mode m at frequency F.

In the presence of dispersive material, that is, assuming that n_1 and n_2 depend
on ω, $dF/d\omega$ is, in general, a function of ω:

$$\frac{1}{(u_z)_m} = C_m(F) \cdot \frac{dF}{d\omega} \tag{4.11}$$

In the unique fundamental mode, if the modulated source has a small spread,
there is no modal dispersion.

For a source of light having a spread in the vicinity of λ_0 an intramodal
dispersion exists for each mode. This exists even for the unique fundamental
mode if the group velocities are not equal at the extremities of the band covered
by the modulated source.*

In summary, spreading of transmission duration, so harmful in communications, is caused:

(1) By modes of the guide which have different transmission times. To avoid this
    intermodal spreading one can:

---

* Variation of the first term $C_0(F)$ can be compensated by variation of the second term $dF/d\omega$,

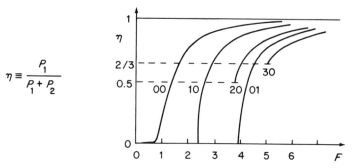

$$\eta \equiv \frac{P_1}{P_1 + P_2}$$

Fig. 4.4. Efficiency of guidance. P_1 is the power in the core and P_2 is the power in the cladding. $F = a(K_1{}^2 - K_2{}^2)^{1/2}$. 00 is the fundamental mode (here, the numbering of the modes is that of Helmholtz's scalar approximation)

(a) Reduce the separation $k_2 - k_1$ (SI multimode fibres with low guidance, $\Delta \ll 1$, allow this result to be obtained);

(b) Operate in the unique fundamental mode (SI monomode fibres are specially produced for this effect).

(2) (Only for a light source having a spread in wavelength) by normal or chromatic intramodal dispersion and characteristic dispersion of the materials of which the guide is made, called 'chromatic material dispersion'.

4.6 POWER IN THE CLADDING

If P_1 and P_2 are the optical powers in the core and cladding, respectively, the efficiency of guidance $P_1/(P_1 + P_2)$ is an increasing function of v at cutoff.

Thus for $v = 1$ and $v = 2$ nearly all the power is in the cladding at cutoff. For $v > 2$ the power in the core is no longer negligible. For all modes, the efficiency of guidance increases for $F > F_c$.

For example, for the fundamental mode and $F = 2$, the guiding efficiency is approximately 0.77 (see Fig. 4.4).

(*Note*: For all modes, radial propagation of energy in the cladding is zero if $F \geqslant F_c$. This suppression of radial propagation is the same condition for guidance without loss. The energy of the wave in the cladding decreases, approximately exponentially, as an increasing function of $r(r > a)$.

5 The inhomogeneous dielectric rod

5.1 INTRODUCTION

The reader will recall that all material used for guidance in a given direction will present a variation of physical characteristics as a function of a transverse co-ordinate. Optical guidance relies on cylinders of either rectangular or circular section which in practice are ribbons or fibres, respectively.

The model of the step index fibre is a dielectric rod surrounded by an early homogeneous external medium. That of the graded index fibre is an inhomogeneous dielectric rod surrounded by an external homogeneous medium. In this rod all points situated on the same section parallel to the $0z$ axis have the same physical characteristics. In contrast, the refractive index varies as a function of the distance r from the $0z$ axis and decreases slowly from the value n_1 for $r = 0$ to the value n_2 for $r \geqslant a$.

The model is a symmetrical cylinder of revolution. The variation of refractive index is continuous and slow and the first derivative is also continuous. Continuity is necessary so that, in the vicinity of any point in the rod, the medium can be considered to be locally homogeneous and isotropic for a distance large compared with the local wavelength λ.

The model considered has a 'power law' index:

$$n(r) = \begin{cases} n_1 \left[1 - 2\Delta \left(\dfrac{r}{a} \right)^g \right]^{1/2} & \text{for } r \leqslant a \\ n_2 & \text{for } r > a \end{cases} \tag{5.1}$$

where r is the distance from the $0z$ axis, $n_1 > n_2$ and $2\Delta = (n_1{}^2 - n_2{}^2)/n_1^2$ (Fig. 5.1). In the particular case where $g = 2$, the power law is called a 'parabolic law'.

In this chapter we shall briefly present use of the eikonal equation in view of its application to the cylindrical lens and the WKB approximation applied to the graded index fibre. These approximations relate to geometrical optics.

54

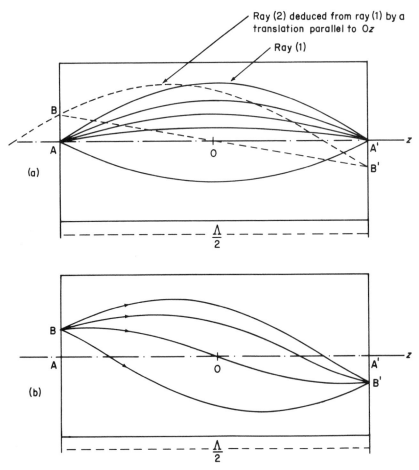

Fig. 5.1. Cylindrical lens: a rod of graded transverse refractive index. (a) The meridional rays, from a point A on the axis of symmetry $0z$ of the cylinder form sinusoidal trajectories in the inhomogeneous medium which cut the $0z$ axis at A'. The distance AA' is a spatial half-wave. (b) In these conditions there is stigmatism for meridional rays between each point A of the input cross-section and its symmetrical point A' with respect to 0, the centre of the cylinder. A sinusoidal trajectory, displaced by a translation parallel to $0z$ is again a tragectory.

For non-meridional rays, that is, helical ones from A, not far from the meridional plane the output points are distributed on an arc of a circle on the cross-sectional plane centred on $0z$ and passing through A'. In practice there is good correspondence between a small area surrounding the point A and a small area surrounding the point A'

5.2 THE EIKONAL EQUATION AND SINUSOIDAL TRAJECTORIES

It is required to determine luminous rays which are the trajectories orthogonal to surfaces of equal phase, called wave surfaces, in an isotropic medium. In the particular case where $g = 2$ the 'eikonal' equation provides the following approximate solution:

$$\frac{r}{a} = U_0 \cdot \cos\left(\sqrt{(2\Delta)} \cdot \frac{z}{a}\right) \tag{5.2}$$

where U_0 is an excursion parameter: $U_0 = (r/a)_{max}$.

Each trajectory has a spatial periodicity of length:

$$\Lambda = a \frac{2\pi}{\sqrt{(2\Delta)}} \tag{5.3}$$

The maximal angle θ_0 of the ray to the $0z$ axis occurs for $r = 0$, and is given by:

$$tg\theta_0 = \sqrt{(2\Delta)} \cdot U_0 \tag{5.4}$$

The meridional trajectories in the rod form sinusoids. (This heuristic solution could be envisaged.) Another trajectory could be deduced from each trajectory by a translation parallel to the axis, or by a rotation about the axis.

If one considers two cross-sectional planes separated by a distance $z_1 - z_2 = \Lambda/2$, one has stigmatism for the meridional rays between any point M_1 on the cross-section $z = z_1$ and the point M_2 on the cross-section $z = z_2$ symmetrical to M_1 with respect to point C of abscissa $(z_1 - z_2)/2$ on the $0z$ axis.

It can also be shown that, for non-meridional rays, there is also approximate stigmatism between M_1 and M_2 on condition that the skew trajectories are not too far from the meridional plane.

In conclusion, if one considers a point M on a ray, the projection m of this point on the $0z$ axis travels with velocity $v = c/n_1$. All optical paths between the point M_1 of abscissa z_1 and the point M_2 of abscissa z_2 are equal and consequently there is equality of the transmission duration at the phase velocity between the input and output cross-sections. In harmonic conditions of angular frequency ω a wavefront perpendicular to $0z$ travels at velocity v_1 in the guide.

If there is a group of waves with a small spread about the angular frequency ω and if the media are dispersive (the refractive index n is a function of r and ω), the group propagates at each radius, at each instant, at the group velocity u, given by:

$$\frac{1}{u} = \frac{1}{c}\left(n + \omega \frac{dn}{d\omega}\right) \tag{5.4}$$

In these conditions, the transmission times from point M_1 to point M_2 will, in general, be unequal.

The numerical aperture on the axis is given by $NA(0) = n_1 \sin\theta_0$, for $U_0 = 1$:

$$NA(0) = n_1\sqrt{(2\Delta)} \tag{5.5}$$

and away from the axis:

$$NA(r/a) = NA(0)\sqrt{\left(1 - \left(\frac{r}{a}\right)^2\right)} \tag{5.6}$$

It becomes zero for $r = a$ and the effective numerical aperture of the input cross-section is:

$$NA_{eff} = NA(0)/\sqrt{2} \tag{5.7}$$

The quantity of light which can be accepted, for guiding, by a graded index rod is half of that which can be accepted by a homogeneous rod, all other things being equal.

5.3 THE WENTZEL, KRAMERS AND BRILLOUIN (WKB) APPROXIMATION

The WKB approximation relates to geometrical optics. It applies well to graded index fibres, but gives little information in the vicinity of cutoff. The light ray within the fibre is curved, generally skew and repetitive, so that it can be reproduced by an elementary rotation $\Delta\phi$ about the $0z$ axis or by an elementary translation Δz along the same axis. A guided ray forms a propagation mode if it superposes on itself by projection on a cross-sectional plane of the fibre to form a stationary wave there.

To determine the periodicity, the azimuthal component of the wave vector $\vec{k}$ must contain an integral number of periods in one circumference of radius r. Similarly, the radial component of $\vec{k}$ must consist of an integral number of half-periods between the 'turning points' where the ray reverses its radial movement. A mode (v, μ), can be represented by a fan of rays performing a helical movement, reflected repetitively on an exterior cylindrical limit of radius r_2 and an interior cylindrical limit of radius r_1. These radii r_1 and r_2 are defined by the condition that the phase of the field has a unique value there.

Hence the ray of a mode defined by two integer parameters is tangential to two caustic curves on which it has the same phase repetitively. In short, the modes correspond to a radial resonance and an azimuthal resonance. In a cylindrical cavity, the resonant modes have three integer parameters; in an infinite rod there are only two.

5.3.1 The number of modes

For a given mode (v, μ), light travels between the interior caustic curve of radius r_1 and the exterior caustic curve of radius r_2. Beyond these limits are regions penetrated by evanescent waves.

The total number of guided modes is obtained from a power law profile by the relation:

$$N = \frac{g}{2(g + 2)}(ak_0n_1)^2 \cdot 2\Delta \tag{5.8}$$

where the normalized frequency defined in connection with the homogeneous rod reappears:

$$F = a(k_1{}^2 - k_2{}^2)^{1/2} \tag{5.9}$$

or

$$F = a \cdot k_0 \cdot n_1 \sqrt{(2\Delta)} \tag{5.10}$$

Hence

$$N = \frac{g}{g+2} \frac{1}{2} F^2 \tag{5.11}$$

For a step index fibre ($g = \infty$) it is found that:

$$N \simeq \tfrac{1}{2} F^2 \tag{5.12}$$

For a parabolically graded fibre ($g = 2$) it is found that:

$$N \simeq \tfrac{1}{4} F^2 \tag{5.13}$$

In consequence, at equal normalized frequencies the fibre with parabolically graded index guides half as many modes, that is, half the light, as the step index fibre, when all modes are excited with the same energy.

5.3.2 The field equation and the axial wave number β

The vector wave equation:

$$\Delta \vec{E} - \varepsilon \mu_0 \frac{\partial^2 \vec{E}}{\partial t^2} = 0$$

allows scalar expressions to be written for the transverse components of the electric field in the form suggested by the results obtained for the homogeneous rod. For example:

$$E_x = \psi_1(r) \begin{pmatrix} \cos v\phi \\ \sin v\phi \end{pmatrix} \cdot e^{-j\beta z} \tag{5.14}$$

where v denotes the integer azimuth parameter. This leads to the expression:

$$\frac{d^2\psi_1}{dr^2} + \frac{1}{r}\frac{d\psi_1}{dr} + \left[k_0^2 n^2 - \beta^2 - \frac{v^2}{r^2} \right] \psi_1 = 0 \tag{5.15}$$

However, the mathematical analysis is greatly simplified if it is assumed that the index profile is not truncated but continues indefinitely beyond the radius $r = a$, following the broken line curve in Fig. 5.2.

Hence, for the parabolic profile structure defined for $g = 2$ by equations (5.1), approximately LP (linearly polarized) modal solutions can be expressed by Laguerre–Gauss functions. LP modes were proposed by Gloge (1971), who noticed that for weak guidance the modes of a homogeneous rod are very nearly linearly polarized and that the field components can be obtained from the principal transverse component of the electric field. In contrast, the singular point at $r = a$ and the discontinuity of the derivative of the function $n(r)$ complicate the mathematical analysis.

In practice, most modes decrease rapidly in the core of the fibre and do not

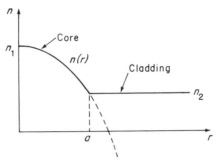

Fig. 5.2. Graph of variation of refractive
index $n(r)$ as a function of distance r from
the $0z$ axis

interact with the cladding to an important or even appreciable extent, so the
assumption of unlimited parabolic decrease can be made. Modes which interact
with the cladding are near to their cutoff and can be neglected, to a first
approximation.

Hence the wave number β of the $LP_{\nu\mu}$ mode is obtained for the fibre with a
parabolic profile index law and for $\Delta \ll 1$ by:

$$\beta = k_0 n_1 - (2\mu + \nu + 1)\frac{\sqrt{(2\Delta)}}{a} \qquad (5.16)$$

Thus the wave numbers are in arithmetic progression of ratio:

$$\frac{\sqrt{(2\Delta)}}{a}$$

5.3.3 The propagation condition for a unique guided mode

Okamoto and Okoshi (1976) presented an analysis of a rod with an index profile
which results in the existence of a single guided mode for $F = 0$, up to:

$$F_0 = 2.405\left(1 + \frac{2}{g}\right)^{1/2} \qquad (5.17)$$

For a parabolic profile $g = 2$ and $F_0 \simeq 3.401$ and for a step index $g = 0$ and
$F_0 = 2.405$.

5.4 MODAL DISPERSION

In most applications fibres are used to transmit a form of digital modulation
and performance is characterized in terms of degradation of the propagated
impulse.

In order to characterize this distorsion the output impulse profile is measured, that is, the optical power as a function of time $h(t)$, for an input pulse of negligible duration.

Normally, the 'impulse response' $h(t)$ is characterized by a unique parameter, the quadratic mean width 2σ:

$$P = \int_{-\infty}^{+\infty} h(t)\,dt \tag{5.18}$$

is the energy of the impulse

$$\tau = \frac{1}{P} \int_{-\infty}^{+\infty} h(t) \cdot t \cdot dt \tag{5.19}$$

is the mean of the reception times and

$$\sigma^2 = \frac{1}{P} \int_{-\infty}^{+\infty} h(t) \cdot t^2 \cdot dt - \tau^2 \tag{5.20}$$

2σ is the quadratic mean width of the impulse.

If it is assumed, as a first approximation, that all guided modes have been excited with the same energy, all modes must be considered from the fundamental to the mode of highest rank defined by the cutoff condition $F_c \leqslant F$. The transit time $\tau(F)$ of the group in this mode can be calculated. There are two cases to be considered, when the profile index has been optimized and when it has not, to reduce the separation between the fundamental and a mode of given higher rank by assuming $\Delta \ll 1$ and taking account of its variation as a function of λ_0. Hence an optimal value g_0 of g can be found. The other case, $g \neq g_0$, is not optimal for the wavelength of the light.

It is found that:

$$\text{for} \quad g \neq g_0 \quad \tau(F) = \Delta \frac{g - g_0}{g + 2} \tag{5.21}$$

$$\text{for} \quad g = g_0 \quad \tau(F) = \Delta^2 \cdot \tfrac{1}{2} \tag{5.22}$$

Figure 5.3 presents power distributions $h(t)$ as a function of time for $g = 1, 4, 10$ and ∞ and also for $g = g_0 = 2$. If $g > g_0$ the modes of higher rank arrive after the fundamental and if $g < g_0$ they arrive before. If $g = g_0$ the power distribution is rectangular as for $g = \infty$, but its magnitude is $\Delta^2/2$ instead of Δ.

Everything of interest can be assessed there for a GI fibre in comparison with a multimode SI fibre; an impulse response which can be much shorter. The ratio is $\Delta/2$, usually about 50. For present-day silica fibres, g_0 is between 1.7 and 2.3.

In practice, the received pulse has a tail due to weak leakage modes. However, the essential characteristic of the pulse is not its total magnitude but its quadratic mean width 2σ. σ has a value of $(12)^{-1/2} = 0.288$ of the total magnitude of a rectangular impulse and a value of $(18)^{-1/2} = 0.236$ of the total magnitude of a triangular impulse. Thus in the case where all the guided modes which leave the fibre have been uniformly excited at the input and uniformly attenuated by

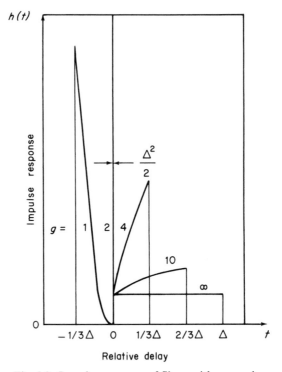

Fig. 5.3. Impulse response of fibres with power law refractive index profile, for different values of exponent g. If $g > 2$, the modes of higher rank arrive after the fundamental. If $g < 2$, they arrive before. For $g = \infty$, the power distribution $h(t)$ is rectangular

the propagation:

$$\text{For the multimode SI fibre} \quad \sigma = 0.288 L N_1 \frac{\Delta^*}{c} \qquad (5.23)$$

$$\text{For the optimized GI fibre} \quad \sigma = 0.144 L N_1 \frac{\Delta^{2*}}{c} \qquad (5.24)$$

The presence of tails on the pulses leads to a replacement of the coefficient 0.288 by 0.388 and 0.144 by 0.150.

For the non-optimized GI fibre, taking account of the tails of the impulse:

$$\sigma = 0.246 L N_1 \frac{g - g_0}{g + 2} \cdot \frac{\Delta}{c}. \qquad (5.25)$$

Arnaud (1975) announced that the index profile which minimizes σ and not the total magnitude of the response is close to the power law with an exponent

* N_1 is the group refractive index $N_1 = c/u_1$, since $n_1 = c/v_1$ and L is the length of the fibre.

$g = g_0 - 2.4\Delta$. One then finds:

$$\sigma_{min} = 0.022LN_1 \frac{\Delta^2}{c} \qquad (5.26)$$

This quadratic deviation represents the theoretical limit which cannot be exceeded.

Since the quadratic magnitude of the pulse is 2σ, the golden rule is to provide an available aperture for the pulse of 4σ; this gives a binary rate $b = 1/(4\sigma)$.

In the above case, the maximum rate is therefore:

$$b = \frac{c}{0.088L \cdot N_1 \cdot \Delta^2} \, (\text{bit/s}) \qquad (5.27)$$

This provides a loss of sensitivity of less than 1 dB.

6 Fibre materials and fabrication

6.1 MATERIALS FOR GUIDING STRUCTURES

The reader will recall that guiding of light over long distances is possible only if a very transparent material is available from which the guide can be fabricated. In general, refraction and absorption are closely linked phenomena. Atttempts to find a refracting but non-absorbent material involve separating them. A qualitative description will be presented here as a first approach.

6.1.1 Suppression of absorption

First, electronic transitions will be eliminated; which means that attention will be restricted to frequencies less than those of ultra-violet ($v \leqslant 9.10^{14}$ Hz approximately). In this vast spectral domain, energy can lose its electromagnetic property by causing movement of material particles through the action of the field which exerts a force on them. These forces are exerted on the magnetic or electric moments in the material.

To refer only to electric moments, the first condition for the suppression of absorption is the elimination of electric monopoles and dipoles, that is, electrons, ions and free radicals.

These effects are particularly important at low frequencies. At high frequencies they become less. Towards 10^{11} or 10^{12} Hz, rotary motion of molecules and translation of monopole ions are diminished by their mass. They cannot follow the vibration of the field because of inertia. However, their movement, although small, is unfortunate and is still perceptible at optical frequencies of 10^{14} Hz and beyond.

A brief examination of the static conductivity σ of materials allows those which might have good optical transparency to be distinguished (see Table 6.1).

Metals which have conductivities of the order of 10^7 (mho/m) are very opaque, although they are far from being perfect conductors. At the other extreme, with a value of static conductivity much smaller than for metals by 26 orders of magnitude, synthetic dielectrics offer excellent chances of transparency.

Table 6.1. Order of magnitude of static electric conductivities

Medium	Static σ (Siemens or mho/m)	
Hot plasmas	Very high	
Metals[a]	Greater than 10^7	
Sea water	4 to 5	
Fresh water (rivers and lakes)	$5\ 10^{-3}$	
Granite	10^{-6}	
Pure water[a]	10^{-6}	
Pure ice[a]	10^{-8}	
Dry basalt	10^{-9}	
Ordinary glass[a]	10^{-12}	
Molten quartz[a]	10^{-17}	Region of
Synthetic silica[a]	10^{-20}	conductivities
Cold gases (low density)	Very small	favourable for
Free space	0	transparency

[a] Synthetic products.

Metallic ions of iron, copper, chrome, etc. and also H^+ and OH^- ions have been eliminated from these artificial materials. The latter arise from dissociation of a water molecule H_2O, which is very corrosive and must be regarded as a pollutant of transparent material.

In order to achieve transparency the material must be as free as possible from polar molecules which cause an absorption band by rotation in the far infra-red and ultra-short wave bands. Symmetrical molecules are preferred in these bands.

The silica molecule SiO_2 is polar and therefore angled. In the solid state it is combined in a crystal; only vibration is possible. In the vitreous state, the order at a small distance is tetrahedral as in the crystal. In this form $[CA_4]$ four large oxygen anions O^{2-} in a tetrahedral formation each share a directed electron* p with a small cation Si^{4+} at the centre of the tetrahedron (Fig. 6.1). Thus, the SiO_2 molecule is formed as a tetrahedron and the tetrahedrons are linked together by normal pressure at the vertices by means of oxygen bonds (π or σ). Each oxygen atom is associated at the same time with two silicon atoms, so that to write $SiO_{4/2}$ would be more precise than SiO_2.

The only possible vibrations are those of the basic tetrahedron and the network of tetrahedra. However, X-ray diffraction spectra show that glass has a highly disordered structure, like liquids. In silica glass the oxygen atoms whether bonding or not, form an irregular network which is the vitreous condition of state.

Nevertheless, in such a disordered structure there are breaks in the mixture, having dimensions of 20–200 Å, where the structure of the molecules is uncertain (there are open bonds).

* In the p orbits of oxygen it forms covalent bonds.

64

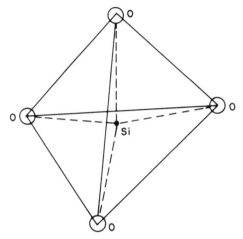

Fig. 6.1. The elementary tetrahedral compaction [CA₄] of silica. A silicon atom links to four atoms of oxygen by covalent bonding, forming a regular tetrahedron

This short introduction to the structure of silica glass shows the extent and complexity of the problem of optical absorption in materials. Theoretical analysis of random networks is essential for an understanding of absorption in these glasses. Research in this direction will permit further explanation of dielectric transparency and progress.

6.1.1.1 Polarizability

Polarization of the orientation of polar molecules is not the only possibility. In fact, the electric field of the em wave acts on the nuclei and causes atomic or ionic polarization by creating an electric dipole moment in the molecule, which is non-polar in the absence of the field. Furthermore, the field acts on the electrons of the atoms and causes electronic polarization by creating another dipole moment.

The effect of atomic polarization is mainly in the infra-red and that of electronic polarization in the ultra-violet. These effects justify Sellmeyer's equation, which is verified by experience:

$$n^2 - 1 = \frac{A_1}{1 - \left(\dfrac{l_1}{\lambda_0}\right)^2} + \frac{A_2}{1 - \left(\dfrac{l_2}{\lambda_0}\right)^2} + \frac{A_3}{1 - \left(\dfrac{l_3}{\lambda_0}\right)^2}$$

The wavelengths l_1, l_2 and l_3 correspond to three resonances of the material due to the effect of the three polarizations. The equation provides typical variation of the refractive index n as a function of the wavelength λ_0 for all polar dielectrics.

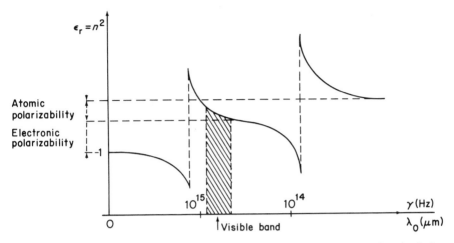

Fig. 6.2. Schematic representation of the variation of the square of the refractive index of a dielectric

If refraction is accurately determined by the phenomena of atomic and electronic polarisation of orientation, absorption in the infra-red region is best determined by the phenomena of molecular vibration–rotation resonance and their harmonics. The resultant absorption lines are broadened by the presence of isotopes.

The essential conclusion is that between the electronic transitions of the ultra-violet and the resonances of the infra-red a band exists between 0.3 and 3 μm or between 10^{15} and 10^{14} Hz where the best chances of dielectric transparency occur (see Fig. 6.2).

6.1.1.2 The particular case of silica

In the infra-red, stretching and bending vibrations of the SiOSi bonds are combined. The fundamental vibration is at 8.3 μm; but the bands combine so that the edge of the infra-red absorption band for dry silica is at 3 μm. However, a water band exists at 2.73 μm, which lowers the practical limit of the transparency band for silica to 2.5 μm.

In the ultra-violet the smallest energy interval between the bands is of 8.9 eV, which roughly fixes the shortest wavelength at 0.14 μm.

Note 1: There are many reasons to consider that glasses such as GeO_2. B_2O_3 and P_2O_5, to cite only a few, could be equally transparent as silica.

Note 2: Residual water in glass creates absorption bands.)

In silica, for an OH content of 1 part per million, the following absorptions are found:

$\lambda_0(\mu m)$	0.95	1.24	1.39	2.73
A(dB/km)	1	2	40	—

The fundamental vibration is at 2.73 μm.

In glass with a very small OH content the bands at 0.95 and 1.24 μm are very attenuated, but pollution can occur.

6.1.2 Scattering and total attenuation

Light is scattered by irregularities or inhomogeneities of the medium where it propagates. In glass used for fibres it is scattered by faults, inclusions, bubbles stretched as the fibre is drawn, mixture breaks, variations in refractive index and, finally, the molecules themselves. In good-quality fibres scattering is principally of the Rayleigh type. The dimensions of the irregularities are less than $\lambda/10$.

Losses by Rayleigh scattering are characterized by an angular dependence of $(1 + \cos^2 \theta)$ and a wavelength dependence of λ^{-4} (see Fig. 6.3).

In defining the exponential attenuation coefficient $e^{-\alpha z}$ three terms are implied, which can be expressed by:

$$\alpha = \frac{a}{\lambda_0^4} + b(\lambda_0) + c$$

where a is the Rayleigh scattering coefficient, b represents the loss variable as a function of λ caused by impurities, including OH, the effect of drawing the fibre and the tails of ultra-violet and infra-red absorption and c is the constant

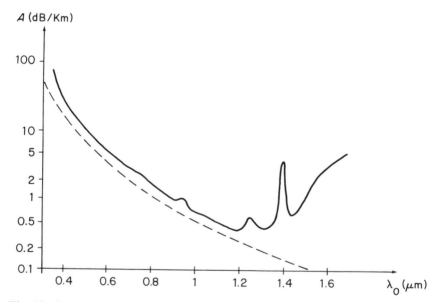

Fig. 6.3. Attenuation A as a function of wavelength in a core of silica doped with germanium. Broken line—the attenuation due to Rayleigh scattering following λ_0^{-4}. Solid line—total attenuation measured experimentally. The difference is principally due to OH content. The attenuation A remains constant between $+70\,°C$ and $-40\,°C$, and increases rapidly when the temperature drops below $-40\,°C$

loss, which depends on input conditions. The first term can be minimized by increasing the operating wavelength and the second may be reduced by purifying the material.

6.2 THE POSSIBLE CHOICES

Guidance of photons in a cylindrical pipe can occur only if the refractive index n_1 of the central material is greater than the index n_2 of the outer tube. Furthermore, the distance covered cannot be large unless the transparency of the two materials is high.

For these conditions, there are a number of solutions:

(1) A cylindrical polycrystalline wire in a cladding of air, inside an absorbant plastic tube;
(2) A capillary glass tube, filled with a more refracting liquid;
(3) A glass wire clad with plastic (or a plastic wire clad with plastic); or
(4) A glass wire, clad with glass.

6.2.1 Solution (3) (core of glass, cladding of plastic)

This solution is readily and widely used, and, is an important improvement over the fibre with a core of glass in a cladding of air. Numerous facilities have been established over the years for the manufacture and use of silica glass fibres clad in plastic.

The interest of these fibres lies in their ease of manufacture in all diameters from 50 μm to 1 mm in long basic lengths at low cost. The best results, particularly those concerning attenuation, have been obtained in France by the Quartz and Silica Company, and important work has been carried out by the central research laboratory of Thomson CSF for the design and production of practical transmission systems using plastic/silica guides in 'terrestrial links'.

Fortunately, there are two types of organic polymer which can be used because they have suitable refractive indices. These are fluorocarbons and silicones. The association of fluorocarbons and silicones with silica allows a large numerical aperture (0.33 to 0.58)* and a small core loss. It is probable that the development of these fibres will involve manufacture of polymers whose chemical structures provide transparency windows corresponding to those of silica. It is required to reduce scattering in these polymers.

Silica fibres clad in plastic are often preferred for shared data transmission or *data busses*, and they resist nuclear radiation well.

Fibres can equally be fabricated in plastic, clad in plastic. These fibres, all plastic, suffer from an insufficient determination of the core–cladding interface and an attenuation of energy in the core greater than that for silicon, so the total attenuation is increased.

Silica–air and silica–plastic fibres benefit from the excellent mechanical and optical properties of silica.

* The *NA* of a silica fibre clad with air is equal to 1.

6.2.2 Solution (4) (core of glass, cladding of glass)

This is the best solution, which provides the highest quality light guides, and it is the most used at present. A glass core covered with a glass cladding is produced and drawn into a fibre together. It is immediately enclosed in a tight-fitting sheath. The two glasses generally have the same composition, to within one dopant. One dopant can be used for the core, to increase the refractive index, and one dopant for the cladding to reduce it.

In the material there are large anions which, because of their polarizability which is fundamentally greater than that of cations, have a very large effect on the speed of photons and hence on the refractive index. For this reason, oxide glasses, mainly composed of large oxygen anions, all have a refractive index around 1.5. To modify the value of refractive index slightly, the mean polarizability of the anions and/or cations is adjusted.

For example, the refractive index is increased by replacing a small percentage of silicon atoms with germanium, from the following row of the periodic table, or particularly with lead.

In certain cases, two dopants are used; one to produce the required deviation Δ and the other, a second parameter, to equalize the group velocities within the spectral bandwidth to be transmitted.

For long-range use over 10–50 km very transparent glass must be used. In general, a total attenuation of not more than 40–60 dB is required between the extremities of the channel.

For use over short distances of several kilometres ordinary glasses can be used. At greater distances propagation is presently dominated by silica SiO_2. Vitreous silica can be obtained from natural crystals and, above all, synthesized chemically from constituents of very great purity. Good homogeneity, excellent transparency and a very good quality fibre can thereby be obtained due to the excellent continuity of the characteristics of this material and the remarkable continuity of the first and second derivatives of these variables.

6.3 GLASS

What is glass? In present terminology, glass:

(1) Is a substance with a wide variety of forms;
(2) Is also a material with a wide variety of compositions;
(3) Is finally a physicochemical state.

The definition currently used by the American Society for Testing Materials is: 'Glass is a mineral product obtained by fusion and solidification after cooling without causing crystallization.' Glass is a solid amorphous mineral, that is, with a disordered structure, in contrast to a monocrystal, whose structure is rigorously ordered and repetitive in three spatial dimensions. It can be considered that in a liquid like water, the molecules (and the macromolecules produced by polymerization) break and re-form their bonds perpetually, thereby giving the material a small viscosity. The viscosity of water is about 10^{-2} poise.

In a vitreous solid the viscosity, in contrast, is very high. In summary, glass is a casting, which is rapidly cooled without nucleation and crystallization having time to occur and is first in a state of superfusion then, from a certain temperature called 'transformation', in a vitreous state. As the temperature decreases, the number of bonds which break and re-form in unit time decreases. Hence the viscosity increases and the glass sets. At ambient temperature, its viscosity becomes 10^{19}–10^{20} poises.

The ratio of the viscosity of glass to that of water is therefore of the order of 10^{22}. When one says that a window is made from a liquid which is slowly running out of its frame it can be seen in what sense this is intended. (The age of the universe is about 3.10^{17} s.) Glasses and mainly silica undoubtedly exist in large quantities in nature. The upper and lower layers of the Earth's crust consist mainly of crystalline silica, or even vitreous silica in the vicinity of the central core, which appears to be surrounded by a material having the properties of a liquid.

At the surface of the Earth's crust, natural glass is rare. However, showers of tectites and microtectites which have fallen on the Earth from space on many occasions during geological times, amounting to billions of tons, are composed largely of drops of molten glass and iron. Blocks of glass are still found (for example, in the Libyan desert) which have probably come from space. Fulgurites are also found which are produced by the effect of lightning on sand, and are strangely shaped objects with elongated branches.

Finally, obsidian, a very hard and resistant mineral which is produced by the heat and pressure of volcanic eruptions, is found on the Earth's surface. Obsidian is opaque and almost always black, sometimes veined with red. Pieces of obsidian have been found everywhere in the world, fashioned by primitive man into objects such as arrow-heads and knives. A splinter of obsidian made by a blow from a pointed stone can present a cutting edge as fine as that of a scalpel made from the best steel.

The many thousands of years which have intervened give an idea of the resistance of glass to corrosion.

6.4 THE PRODUCTION OF GLASS

Glass offers an immense variety of compositions which change its physical characteristics, and the major companies have produced several tens of thousands of kinds of glass.

Ordinary glass consists of 75% SiO_2, 15% NaO_2, 9% CaO and 1% assorted. By replacing CaO with PbO a brilliant glass is obtained which is renowned for its beauty and is malleable when hot, easy to work and very solid after cooling. This heavy glass is used for objects of art, crystalware and television screens since it absorbs X-rays well.

Glass is very sensitive to the inclusion of metallic oxides. For example, nickel oxide in a proportion of 1/50 000 colours glasses between clear yellow and purple, depending on their initial composition, and cobalt in a proportion of 1/10 000 colours glass an intense blue. Flame-red glass is obtained with traces of gold, copper or selenium oxide.

It is necessary to take serious precautions in order to obtain glass which is colourless, or reputedly so, because if glass is observed in a slab, it invariably shows a colour, usually green, even for a single metre of light travel.

The technology of fibres goes well beyond these results, since it allows the production of glasses with kilometric attenuation of several tenths of a decibel; for example, 0.3 dB/km. Hence a fibre of this quality, suspended vertically in a depth of 11 000 m of ocean water from the bottom of the Marianne Trough, allows transmission of light λ_0, captured by its entrance pupil, with only 55% loss.

6.5 FABRICATION OF OPTICAL FIBRES

6.5.1 Glass-making methods

Among these, the double-crucible method is generally used. The starting material is a formative such as silica SiO_2 or anhydrous vitreous borax B_2O_3. A modifier or simply a dopant may be added. The two crucibles are raised to a suitable temperature and a fibre is drawn (see Fig. 6.4).

The interior crucible provides the glass for the core and the exterior one the cladding. It is possible to obtain a thin zone of graded index by ionic diffusion in the region where the two glasses are in contact at the output of the crucibles (Selfoc fibres).

At high temperature the glass can be drawn out like a thread of oil or golden syrup. Iron is malleable only within a narrow temperature interval, then it melts at 100 °C. It is the marvellous thermodynamic continuity of glass which allows a simple tube to be raised to a suitable temperature and drawn out semi-indefinitely. The diameter reduces continuously and the cylindrical tube remains very nearly cylindrical, with only some percentage of ellipticity.

Manufacturers exploit the natural properties of glass to the full, and have knowledge and skills which assure their success.

6.5.2 Modern methods

Instead of using a natural crystalline mineral, such as quartz, to obtain the silica glass chemical substances are used. These are already used in the semiconductor industry for the synthesis of vitreous silica and other vitreous oxides, such as GeO_2, which generally serve as dopants.

The method is of value for numerous oxide glasses, and it is currently used above all for silica, which gives excellent results. The essential fact is that the well-known viscosity of silica is maintained at a high level at high temperatures and silica vapour is obtained in the presence of the viscous melt. Hence silica evaporates totally before a liquid can be obtained. In the inverse operation, which would normally be called a deposit as of snow or soot, the silica vapour

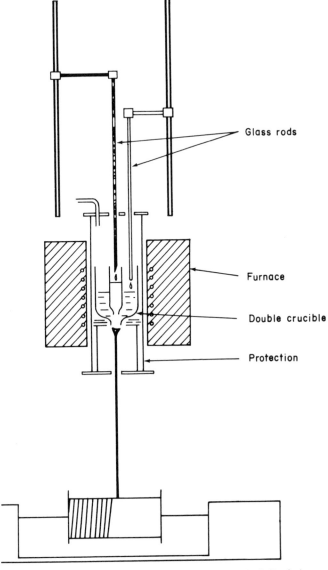

Fig. 6.4. The double-crucible method (after K. J. Beales)

is deposited on cool inner walls in layers which can be homogenized and made uniform.

Table 6.2 allows a comparison of rock crystal glass obtained by melting crystobalite at 1710 °C and synthetic silica glass to be made.

Although apparently a little less dense than crystobalite glass, synthetic silica glass is very homogeneous, very isotropic and very pure. It remains to obtain

72

Table 6.2

Characteristic	Unit	Rock crystal glass	Synthetic silica glass
Density	g/cm^3	2.203	2.201
Working temperature	°C	1,700–2,100	1,600–2,000
Maximum continuous temperature	°C	1,100	950
Refractive index in H_α	—	1.456 46	1.456 37
Static conductivity (σ)	mho/m	10^{-18}	10^{-20}

silica vapour from initially purified material; and to add, if necessary, the vapour of the dopant in order to produce fibres having a given refractive index profile which can vary within wide limits.

6.5.2.1 Discontinuous fabrication

A 'preform' is produced first, that is a, rod of one or two centimetres diameter and a maximum of 1 m long. Thus a certain volume of glass is provided which enables a number of kilometres of fibre to be produced, of which the cross-section, after drawing, will be approximately a scaling down of the cross-section of the preform. Since the thermal conductivities and viscosities are not the same, and because of ionic diffusion, there will be some difference between the profile of the preform and that of the fibre. However, the manufacturer's knowledge allows the desired profile to be obtained to a good approximation, with rare exceptions where the curves representing the variation of the refractive index parameter as a function of dopant concentration have singular points.

The advantage of the discontinuous system is the ability to stock base material conveniently which can eventually produce fibres whose diameters, lengths and packaging are not necessarily fixed in advance.

To produce glass preforms of excellent quality, one starts by purifying the chemical substances to be used; these are halogens in the liquid phase which can be distilled.

Having achieved this purification, a suitable mixture is produced and these vapours are oxidized or hydrolysed in a flame to produce a deposit of the required mixture.

In order to obtain pure silica from silicon tetrachloride,

$$SiCl_4 + O_2 \rightarrow SiO_2 + 2Cl_2$$

a mixture of chloride, pure oxygen and combustible gas are fed to the burner. A deposit is thus obtained on the lining and the chlorine gas is drawn off.

The deposit of glass, formed as a porous layer on the lining, is then heated until vitrification is complete and the porosity disappears. A thin layer of pure silica is finally obtained on the lining (see Fig. 6.5). By modifying the

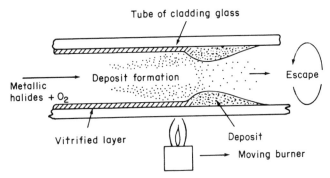

Fig. 6.5. IVPO process: internal vapour phase oxidation. By varying the composition of the mixture of input halogens, the refractive index of the vitrified layer can be varied in order to produce a preform with graded index

concentration of one or more dopants a step index or a series of small changes can be made. This enables the production of 60, 500 or 1000 successive homogeneous layers which are partially smoothed by diffusion of the dopants between adjacent layers.

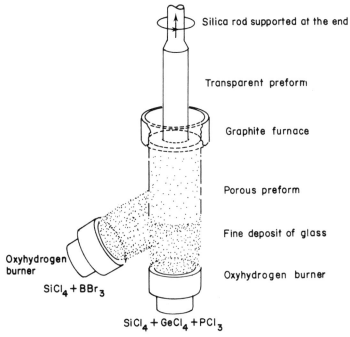

Fig. 6.6. EVPO process: external vapour phase oxidation. The lower burner produces the core glass and the side burner the cladding glass. The preform emerges from the top of the vitrification oven. In principle, its length is unlimited

74

6.5.2.2 Continuous fabrication

In this technique the preform increases in the axial direction. Having made an axial deposit of the core glass, the cladding glass is deposited laterally and the whole is vitrified in a furnace. In principle, production can continue without limit (see Fig. 6.6).

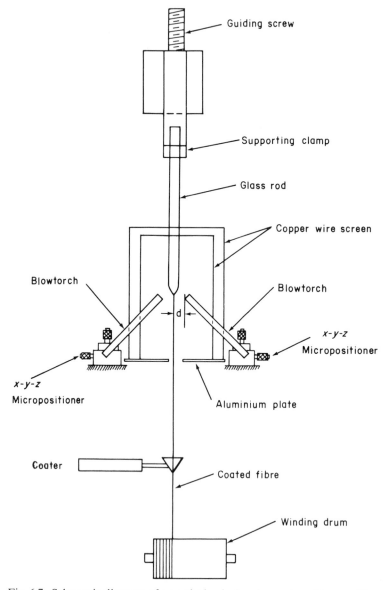

Fig. 6.7. Schematic diagram of a vertical axis drawing system. Drawn silica fibres, prepared in the oxyhydrogen flame

6.5.2.3 Fibre drawing

The double-crucible method allows fibres to be drawn directly. Use of a preform takes fibre production back to a further stage of fabrication (see Fig. 6.7).

Having extracted the preform from its impervious package it is placed on a positioning device, washed with acetone, passed through the flame of a blowtorch and drawn in a controlled atmosphere. The silica is drawn and immediately packaged. As the capacity of reels is limited, a system of pulleys, similar to a multiple pulley block, is used to store the fibre while one reel is replaced with an empty one without interrupting drawing.

6.5.2.4 Packaging

The fibre which has been drawn has a mechanical resistance which approaches the theoretical value of cohesion and is high. However, its mechanical resistance, as with all materials, depends very much on the state of the surface. This must therefore be protected immediately against abrasion, chemical corrosion, humidity and force.

EVA (ethylene vinyl acetate), TEFLON (a fluorocarbon resin) and HYTREL (a polyester elastomer) are often used and applied online by extrusion.

(Note 1: To remove the covering from a fibre it is plunged into a bath of propylene glycol at 160 °C for about 30 s. The fibre then passes between clamps covered with soft polyurethane foam ('Scott felt') which retains the covering without abrasion.

Note 2: To remove the vitreous cladding of a fibre it can be plunged into a 50% solution of hydrofluoric acid (HF) or a saturated solution of ammonium bifluoride (ABF). ABF removes 15 μm in 20 min and HF removes 30 μm in 30 min.)

7 Optical cables and their connection

7.1 DIVERSITY OF USES

The optical fibre is a thin stand of glass which is protected by a sheath. In this form, the fibre is still unsuitable for use, and it must be further protected to prevent pollution by water and mechanical damage. A single-fibre cable can be very thin, light and resistant. Multiple fibre cables are also made (up to 100 fibres or more).

The structure of an optical cable depends essentially on the conditions of use: for example, overhead, underwater, underground; fixed, special for aircraft, special for ships; all-terrain, laid by helicopter; special for a railway, tunnel or mine gallery conduit, etc.

The basic length of the cable depends on the type of fibre, its quality and intended use. Certain cables are combined optical–electrical and include supply conductors with the optical fibres for one or more repeaters or regenerators.

The main quality of a cable is to resist mechanical forces during and after installation in order to suppress or minimize those forces which are imposed on the optical fibre.

7.2 THE CLASSICAL STRUCTURE (Fig. 7.1)

Cable manufacturers have applied their traditional methods to optical fibres. For example, the fibres are assembled around a central carrier, each protected by a polyurethane tube. This is surrounded by a polythene sheath, an aluminium band and finally teflon. Consequently the fibre is very well protected for pulling through a conduct or underground installation, and its mechanical strength is high in tension and compression.

The techniques and equipment required for fabrication are available. However, the bulk and weight of a drum of cable of a given length appear to be too large. The cost is high and joining fibres is complicated.

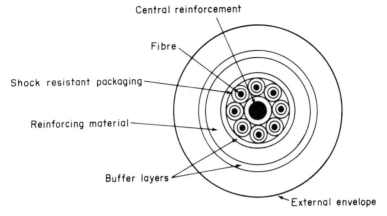

Central reinforcement

Fibre

Shock resistant packaging

Reinforcing material

Buffer layers

External envelope

Fig. 7.1. Optical cable with a compartmented cylindrical structure

7.3 RIBBON STRUCTURE (Fig. 7.2)

The first cables having this structure were made by Bell Laboratories, and they allow a very large density of fibres (for example, 144) by using 12 ribbons each of 12 fibres for a diameter of 12 mm. (A ribbon of plastic material has 12 longitudinal slots each containing one fibre.) The assembly is enclosed helically in a strong sheath. The advantage of such a structure is the possibility of a global link having a large number of fibres.

There are also cables of the same class which use ribbons of corrugated aluminium. A stack of ribbons provides channels in which the fibres are placed.

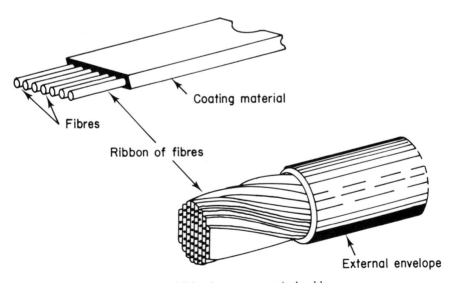

Coating material

Fibres

Ribbon of fibres

External envelope

Fig. 7.2. 'Ribbon' structure optical cable

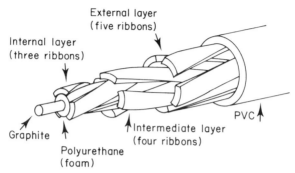

Fig. 7.3. Cable with helical ribbon structure. Tension is
reduced by diametric contraction

7.4 HELICAL CHANNEL STRUCTURE

A cylindrical carrier containing helical channels is placed around a central
reinforcement which is intended to resist traction applied to the cable. The
fibres are deposited in the channels without tension and surrounded by a
polythene foam. This carrier is wrapped in a soft ribbon and the whole is
enclosed in a sheath which resists crushing.

The helical channel structure fortunately combines the advantages of the two
preceding structures—reduced bulk and simultaneous protection of the fibres.
If a tractive force is applied to the cable it stretches and the longitudinal
reinforcements resist this tractive force in such a way that the diameter of the
helical structure decreases slightly without changing the curvilinear length and
without applying appreciable tension to the fibres (see Fig. 7.3).

(*Note*: The unit of constraint is the Newton per square metre. The English
unit is the psi; $1 \ k$ psi $= 10^3$ psi $\simeq 6.9 \ 10^6$ Pascal or N/m^2.)

7.5 JUNCTIONS (SPLICES AND CONNECTIONS)

Two types of junction can be distinguished: those which are permanent are
made with splices and those which may be disconnected and reconnected a
number of times are made with connectors.

Certain splices must be made in the factory before or during cable
manufacture. This is particularly the case when the length of fibre which can
be made from the preform is less than the basic length of continuous cable
required. For example, to produce a basic cable length of 21 km it may be
necessary to splice three fibre segments each of 7 km. The two necessary splices
are made on the ends of the bare fibres using an electric arc, under the microscope;
this causes a very short length of glass to melt for a short time. (Fig. 7.4). Having
made two 21 km cables, these are in turn spliced in the field in order to obtain
a 42 km length of optical cable.

Other splices are made in a capillary tube where the interconnection of the

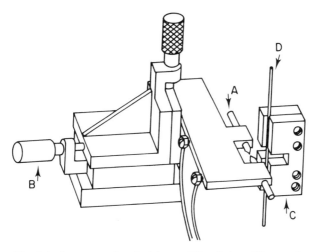

Fig. 7.4. Apparatus for electric arc (A) splicing of two end-
to-end fibres (D)

optical media is assured by filling the tube with an epoxy resin of matching
refractive index (Fig. 7.5).

A good splice has a mean loss of 0.1–0.15 dB, according to the method used.
A good joint using connectors has a mean loss of 0.5 dB. Consequently a large
number of connectors cannot be used; they are, however, necessary.

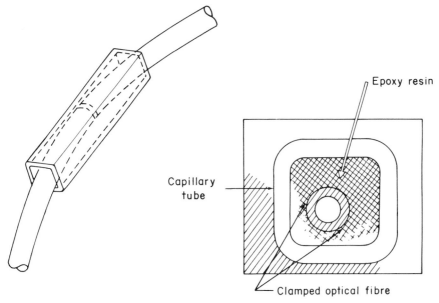

Epoxy resin

Capillary
tube

Clamped optical fibre

Fig. 7.5. Splicing in a square section capillary tube. A stream of epoxy resin ensures
both immobility of the extremities of the fibres and optical matching

80

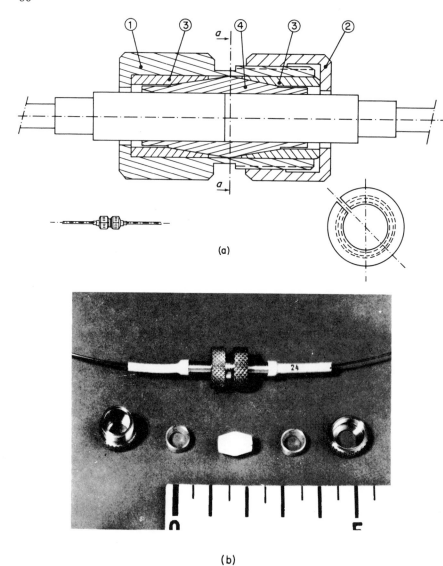

(a)

(b)

Fig. 7.6. Two collars (3) with a conical taper identical to that of the sleeve adapt themselves to it. Their role is to modify the diameter of the sleeve by tightening as they approach while preserving the alignment[1]

Fortunately, for a given capacity it is possible to produce sections of optical cable which are much longer than sections of metallic cable of equal weight or bulk.

In joining optical fibres using connectors difficulties arise from the small dimensions and tight tolerances. For a connection of two optical fibres of $NA = 0.2$ and diameter $2a = 50\,\mu m$, it is necessary:

(1) To polish the extremities with two or three abrasives;
(2) To locate the end sections within 10 μm of each other;
(3) To locate the axes within 3 μm of each other; and
(4) To align the axes within 2 degrees.

This enables a mean connection attenuation of 0.5 dB to be obtained with the inclusion of a liquid of matching refractive index.

There are several techniques, all in competition with each other, and a tendency towards miniaturization is evident. There is, however, a lack of standardization of dimensions (Fig. 7.6).

8 Sources of light for optical guidance

8.1 GENERAL

To be useful at a certain quality level a received light signal must satisfy a number of conditions, among which it must:

(1) Remain above a certain power threshold; and
(2) Occupy a minimum bandwidth.

For the moment, consider the minimum power threshold. It is clear that to increase the range, that is, the distance between the transmitter and receiver, it is desirable to maximize the power entering the fibre and consequently the power of the source. There is, however, a limit which cannot be exceeded. The power per unit area injected into the fibre must remain below the threshold above which the first non-linear effect begins to appear. For silica, which is a very linear material, the stimulated Brillouin scattering threshold occurs between 20 and 40 kW/cm^2. Present sources and silica fibres clad with silica are about two orders of magnitude below this limit. Light from several sources can therefore be injected into the same fibre without affecting guidance; different information can be transmitted simultaneously by using several wavelengths. This process is called 'wavelength multiplexing'.

8.1.1 General requirements

In digital modulation, which is less demanding than analogue modulation, the light impulse necessary for reception represents a train of around 100 photons and a transmitted impulse contains 10^5–10^7 photons, the maximum permissible transmission attenuation between the ends of the link being 30–50 dB. (One would like it to be greater.)

For continuous sources intended for analogue amplitude modulation, very linear characteristics are required which do not create undesired harmonics. For pulse sources, intended for digital modulation, very short rise times are required—a few tens of nanoseconds.

The ideal would be a power source of very small dimensions, having a small radiating area and a narrow radiated beam. It would also have both a very brief rise time for pulse mode operation and a very linear characteristic for amplitude modulation over a wide dynamic range.

8.1.2. Particle physics and light source research

The small luminous pencil which enters a light guide can be considered as a collection of photons. Their emission or absorption can be caused by the movement of certain objects, such as atoms, which collide with each other and finally lead to a change of state of electrically charged particles involving electrons.

Furthermore, physics explains, that photons, being bosons, can be created or destroyed, one by one. As far as fibre technology is concerned, several tens of photons, at least, are required in order to characterize the presence of light at the receiver in a small time Δt.

Experience shows that to produce any particle it is necessary to provide energy and cause collisions between other particles in the course of which the desired particle has a chance of appearing. When it appears, it is necessary to optimize the equipment to increase the yield and to satisfy the requirements better.

The electric arc lamp or incandescent filament produce photons by collision. The emission is incoherent with a wide spectral bandwidth. These sources are not applicable to optical fibres because it is not physically possible to position the end of a strand of glass close to an electric arc or filament without melting it.

Consider the use of a 6 V 5 A electric bulb of 3 cm diameter. The area of the glass bulb will be 2.83×10^{-3} m^2 through which an optical power of around $(6 \times 5)/4 = 7.5$ W will flow. In these conditions the optical power entering a fibre of area 2.5×10^{-9} m^2 applied to the glass bulb will be only 7 μW of light, which covers a wide spectral bandwidth.

Such an arrangement is of little use. It could be used in a laboratory by including a monochromator and a condenser but it is not useful for transmission of signals. At present, the most interesting practical devices are:

(1) Electroluminescent emitters, that is, semiconductor diodes which emit in the infra-red when an electric current is applied; and
(2) Solid state miniature lasers.

8.2 SEMICONDUCTOR DIODES

The history of these diodes started in 1962, with gallium arsenide (GaAs) junction diodes at cryogenic temperatures. Stimulated emission was progressively achieved by extending the wavelengths into the visible and near infra-red by exploitation of numerous group III–V and II–VI compounds.

Since 1968 the spectral region of wavelengths from 0.5–20 μm has been covered, although problems of operation at ordinary temperatures have been resolved only for some materials. The most important compound is GaAs, which emits at around 850 nm. The following are distinguished:

(1) Simple light-emitting diodes (LED) which are incoherent emitters; and
(2) Semiconductor laser diodes (SCL), also called 'injection lasers', which are coherent emitters.

These lasers and simple diodes have similar structures, and the principal difference lies in the existence of a Fabry–Perot resonant cavity in the lasers. Otherwise, the structures are very comparable.

In the two cases, emission of light is due to the radiative recombination of electron–hole pairs. The hole is an unoccupied state in the valence band (VB) of the semiconductor, and the transition occurs when an electron returns from the conduction band (CB) to the VB by crossing the forbidden band (FB) of energy levels.

If the recombination is radiative and direct, the photon produced has the energy difference between the two discrete levels:

$$E_2 - E_1 = h\nu$$

Hence the width of the forbidden band determines approximately the frequency ν of the emitted photon.

In semiconductors described as 'indirect', direct recombination between the minimum of the CB and the maximum of the VB is not possible. Intervention of a third particle, the 'phonon', is necessary whose energy is propagated by a mode of vibration of the crystal. The transition is of second order and much less probable than a direct transition. In indirect semiconductors (Si, Ge, AlAs) good internal efficiency is difficult to obtain.

8.2.1 Injection of carriers into a p–n junction

Injection of minority carriers into a forward-biased p–n junction produces a population reversal and causes spontaneous and stimulated emission. Use of heterojunctions, that is, junctions uniting two different semiconductor compounds, permits carrier injection and emission of both spontaneous and stimulated light.

In practice, an active layer is interposed, as a sandwich, between two external layers; this confines the charge carriers to the active layer on account of their larger forbidden band and, if necessary, confines photons on account of their much smaller refractive index. This device constitutes a double heterojunction (DH).

With laser diodes, the active layer forms a Fabry–Pérot resonant cavity having two opposed reflecting faces, obtained by cleavage of a single crystal, and two dull lateral faces obtained by sawing. This cavity produces a gain which exceeds the losses when the current exceeds a certain threshold. Consequently, only stimulated emission remains and provides light, in a number of discrete modes of resonance, if the threshold is exceeded. Emission occurs laterally from the two cleaved faces, and one half is therefore lost to the fibre.

With simple light-emitting diodes the active layer does not form a resonant cavity. There is no laser effect and the emission is incoherent. If the photons are principally confined to the active layer by the refractive indices of neighbouring layers emission occurs laterally.

Other devices favour perpendicular emission of light from the active layer and junctions. Certain of these LEDs have an extended radiating area and can be applied to bundles of fibres illuminated in parallel. Others such as the Burrus diode provide a high luminance in a small area and are applicable to a single fibre.

8.2.2 Current–light transfer characteristics

Figure 8.1 shows the optical power characteristics as a function of injected current for electroluminescent emitters. They allow their output to be assessed. LEDs have a very linear characteristic and are consequently well suited to analogue applications. The quantum efficiency varies by 1% for good planar diodes and several per cent for Burrus diodes.

SCLs have a cranked characteristic, of which the first part, for low intensities, is a mediocre LED characteristic. Above a certain intensity, called the 'threshold current', stimulated luminescence occurs and the quantum efficiency exceeds 30% or 60% for the two faces of the cavity. The second part of the SCL characteristic has a steep slope of mediocre linearity because the different modes

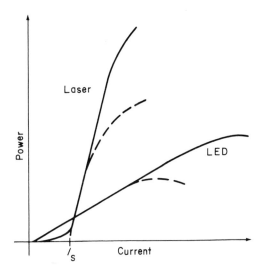

Fig. 8.1. General form of the power characteristic as a function of current, for a light-emitting diode and a laser diode in continuous (— — —) and pulsed (———) operation. (After B. de Crémoux[1])

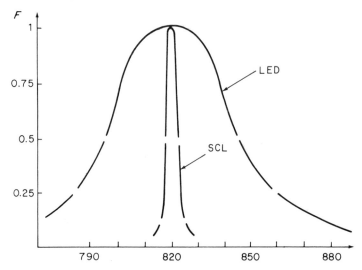

Fig. 8.2. Relative spectral distribution of radiated energy for a laser
diode (SCL) and a light-emitting diode (LED) at 820 nm[1]

of resonance of the cavity are successively stimulated as the current increases;
this produces small but annoying 'kinks' in the practical characteristic which
make the component unsuitable for amplitude modulation (Fig. 8.2).

In contrast, pulse lasers operating with currents of high crest factor are widely
used for digital modulation.

A saturation can be established which corresponds to an upper power limit
for each mode of resonance of the cavity which is more rapidly attained in
continuous than pulsed operation.

8.2.3 Spectral characteristics

The mean wavelength of electroluminescent emission depends on the
material of the semiconductor used. However, the spectral width of the source
depends greatly on the form of emission.

If the emission is entirely spontaneous, the spectral width at 300 °K is,
approximately, 300 Å at 850 nm and 800 Å at 1200 nm.

If the emission is stimulated, just above the threshold (10% above), to excite
only the first mode of the cavity, the spectral width can be very small (0.1 Å).
If the current is further increased, the spectrum can contain several modes and
can reach a width of 30 Å at 3 dB (see Fig. 8.3).

8.2.4 Materials for electroluminescent emitters

Numerous intermetallic compounds have properties similar to those of
germanium and silicon, the simple semiconductors. These elements, which

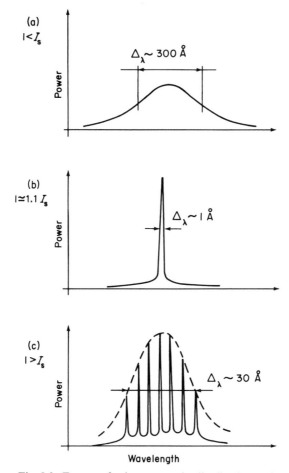

Fig. 8.3. Form of the spectral distribution of radiated energy for a laser diode when the current is (a) less than the threshold I_s, (b) greater than I_s by about 10% and (c) much greater than I_s. (After B. de Crémoux[1])

belong to group IV, have a tetrahedral crystal structure in the form of a centred cubic lattice.

In the external layer of each atom each of the four valence electrons shares one state with a valence electron of an adjacent atom; the two electrons are thus associated with antiparallel spins. Each atom is at the centre of a tetrahedron, and the apexes are occupied by its four closest neighbours which are identical to it. The lattice therefore is face-centred cubic.

The band gap of diamond is very large (7 eV) and that of grey tin is small (0.08 eV) (see Table 10.1).

There is interest in binary compounds which are susceptible to covalent bonding and in which the sum of the valence electrons of the two different atoms of the compound is always 8. Hence, on average, each atom at the centre of a tetrahedron will have four valence electrons with which to form bonds with its four closest neighbours at the apexes of the tetrahedron. This time, the atom at the centre and the atoms at the apexes are different. The process exists for sufficient time for sharing to occur, but the compounds must exist.

It is possible to assign a covalent radius to each atom. The sum of the two radii, in binary compounds, gives an estimate of the lattice parameter and allows the possibilities of substitution and introduction of dopants into an alloy to be assessed. This enables the dimensions of the lattice to be conserved and matching crystalline layers to be superposed. This matching is necessary for operation of electroluminescent diodes over a long period.

Numerous III–V, II–VI and I–VII compounds are therefore used which have the sphalerite crystal structure which is similar to that of diamond. In GaAs, for example, the six face centres in the cubic lattice, the four major apexes and the four minor apexes are occupied by As atoms in such a way that in the two orthogonal diagonal planes the interior atoms are Ga.

Consider substitutions. For example, 8% of gallium atoms are replaced by aluminium atoms. This gives a ternary compound:

$$Al_{0.08} Ga_{0.92} As$$

In indium arsenide, $100x\%$ of indium atoms are replaced by gallium atoms and $100y\%$ of arsenic atoms by phosphorous atoms. This gives a quaternary compound:

$$Ga_x In_{1-x} As_y P_{1-y} \quad (x < 1 \quad \text{and} \quad y < 1)$$

If the resulting substance is a solid, it is described as a 'solid solution'.

By use of these mixtures the size of the forbidden band can be adjusted to cover a gap in the spectrum. Furthermore, quaternary mixtures permit matching of lattices within certain limits.

8.2.5 Fabrication technology

The technique is that of epitaxy in the liquid phase in order to grow layers of different composition and dopage on a substrate of poor regularity. In order to fix dislocations of the crystal lattice in the vicinity of the substrate a buffer layer is interposed, then the lower confining layer is produced, and so on, while observing the match of the lattice. The size of emissive region in the active layer is limited by means of a covering metal ribbon which is narrower than the layer and limits the current and noticeably reduces heating.

8.3 PRACTICAL USES

Electroluminescent sources are now obtainable which have excellent stability at ordinary temperatures and operate continuously for several years whose shape and configuration match those of the fibre.

These sources include LEDs and SCLs, can be directly modulated by current injection and can operate in the best transmission windows of present fibres. The minimum power is several μW and a suitable mean power is from one to several milliwatts. Other methods will probably appear which can be added to these.

8.3.1 Light-emitting diodes

Spontaneous emission normally occurs in the p layers, and non-radiating processes limit the 'quantum efficiency' η. Certain electron–hole pairs do not produce the desired photon $h\nu$:

In a simple homojunction structure: $\eta > 50\%$
In a double heterojunction structure: $60\% < \eta < 80\%$

It is necessary to facilitate extraction of heat. The output power reduces by 2–3 dB when the junction temperature rises by 100 °C.

8.3.1.1 Normal emission from the junction (see Fig. 8.4)

(1) The source is a disc of diameter between 15 and 100 μm (exceptionally, 460 μm with a power of 10 mW);
(2) The thickness of the active region is 10–15 μm;
(3) The radiation is approximately Lambertian.

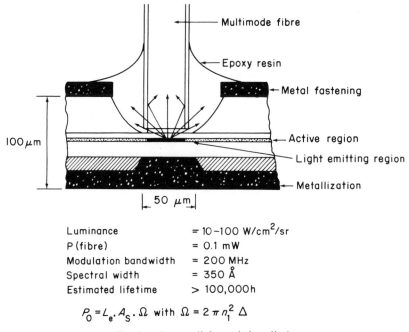

Luminance	= 10–100 W/cm²/sr
P (fibre)	= 0.1 mW
Modulation bandwidth	= 200 MHz
Spectral width	= 350 Å
Estimated lifetime	> 100,000h

$$P_0 = L_e \cdot A_s \cdot \Omega \text{ with } \Omega = 2\pi n_1^2 \Delta$$

Fig. 8.4. Burrus light-emitting diode

90

Example: A normal emission LED at 800 nm and $\theta_s = 75\,\mu m$ is applied to the core of a graded index fibre of $\theta_c = 115\,\mu m$. The radiated power is 15 mW and the power coupled to the core is 1 mW. The current density is 3.4 kA/cm^2, which is half the saturation current.

8.3.1.2 Lateral emission

(1) The radiation is emitted in a narrow astigmatic bundle;
(2) The luminance is high (1000–1500 W/cm^2/sr) and the area of emission is 2–4×10^{-6} cm^2;
(3) The 3 dB spectral bandwidth is in the first window (700–900nm: 250–400 Å) and in the second window (1000–1300 nm: 500–1000 Å).

The spectral width can be reduced only by cooling, which reduces the thermal distribution. The maximum power is fixed by the temperature, but the temperature itself depends very much on the structure. Small devices are cooler.

(*Note*: There are many fabrication difficulties: forward-biased devices for emission are, and will remain, easier to produce than reverse-biased devices for detection.)

8.3.2 Injection lasers SCL (Fig. 8.5)

The laser effect appears only at current densities sufficiently high for the rate of stimulated emission to be greater than the rate of absorption.

Emitted power: For a current of 150–200 mA at 1.6 V (one-third of the power of an ordinary flash) the optical output power is 5 mW per face. It can reach 10 mW.

Lifetime: Degradation of the cleaved faces occurs for an optical power greater than 5 MW/cm^2. Semiconductor lasers (DH) of 1 mW per face can be made which have a mean probable life of approximately 10^6 h. (A year is approximately

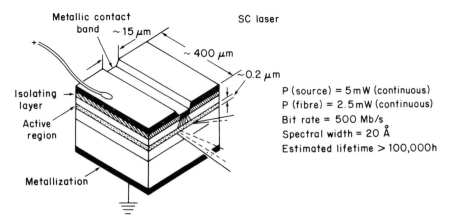

Fig. 8.5. Injection laser

8 800 h.) Good frequency stability of the emission is necessary for wavelength multiplexing.

Parasitic oscillations which are encountered with all oscillators have been progressively eliminated. Here, relaxation oscillations are found around 1 GHz and self-oscillations at several hundreds of MHz:

(1) 3 dB spectral width: 1 Å in longitudinal monomode;
(2) Emissive surface area from 10^{-6} to 10^{-5} cm^2;
(3) The threshold current is less than 4 kA/cm^2.

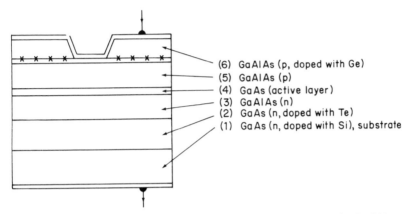

Fig. 8.6. Schematic representation of a double heterojunction laser in the 700 to 950 nm band

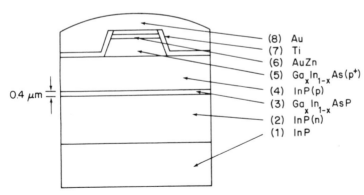

Fig. 8.7. Ga In As P/In P laser at 1300 or 1550 nm. Layers 7 and 8 cover not only the mesa layers (5 and 6) but also the confining layer above 4 from the exterior. In a region of size comparable with that of the mesa layer, the conditions for continuous laser operation at ambient temperature are realized. Threshold currents of 130 mA are obtained for ribbons of 20 μm width and 180 μm length and an active layer thickness of 0.4 μm or 50 mA with a thickness of 0.2 μm. These lasers are transverse monomode types.
(After CNET—Centre Paris B)

Table 8.1. Source characteristics of LEDs and SCLs

Source	λ_0(nm)	Spectral width (Å)	Current (mA)[a]	Dimensions	Width of beam[b]	Optical power (mW)	Luminance (W/cm²/sr)	Rise time (ns)
LED	900	300	300	$\phi = 5$ mm	0.84 sr	1.5	0.90	30
Strong lumin. LED	830	400	150	$\phi = 75\ \mu m$	π sr	10	66	7
Low threshold multimode SCL	820	20	75	$2 \times 13\ \mu m$	$5°_\parallel \times 20°_\perp$	5	—	1
Trans monomode SCL	{810 860}	10	70	$0.2 \times 7\ \mu m$	$10°_\parallel \times 35°_\perp$	7	—	0.1
Longit. monomode SCL	{810 860}	1	70	$0.2 \times 7\ \mu m$	$10°_\parallel \times 35°_\perp$	7	—	0.1
Monomode SCL	{1300 1550}		130	$0.4 \times 20\ \mu m$				
Trans monomode SCL	{1210 1250}		500	$0.2 \times 15\ \mu m$		17	—	
Longit. monomode SCL	{1000 1600}	10	200	$0.3 \times 120\ \mu m$	$6°_\parallel \times 45°_\perp$	4	—	0.75

[a]Control current for LEDs; threshold current for SCLs.
[b]For SCLs, $\parallel$ = parallel to the surface of the junction; $\perp$ perpendicular, half-angle

Modulation can reach the GHz region. The rise times of a pulse are about 1 ns in multimode and 0.1 ns in monomode.

Among the variety of semiconductor alloys used, the following should be particularly noted (see Figs 8.6 and 8.7):

(1) Ga As/Al Ga As in the band 700–950 nm, in DH form:

$$Al_xGa_{1-x}A_s$$

The lattice constant a_0 is different for GaAs and AlAs, but this difference is less than 0.14%. a_0 varies linearly as a function of the variable x of the composition of the alloy. The difference between the lattice constant of GaAs and that of AlGaAs can be appreciably less than 0.14%.

(2) GaInAsP/InP in the band 950–1700 nm, in DH form:

$$Ga_xIn_{1-x}As_yP_{1-y}$$

The fundamental advantage of the quaternary alloy GaInAsP is the possibility of an exact match of its lattice with the supporting InP. The composition can vary from InP to $Ga_{0.465}In_{0.535}As$ with the same lattice constant with only a variation of the band covered. In the band 1210–1250 nm, 15 mW per face are obtained in continuous operation at 20% above the threshold. It can operate in pulse mode at eight times the threshold (Fig. 8.8; see also Table 8.1).

8.4 SOLID STATE LASERS USING 3^+ IONS

8.4.1 General

In lasers, light emitted from one centre in the luminescent material generally stimulates other centres in turn. It is necessary to cause a population inversion, that is, to create a situation where excited states are more numerous than ground

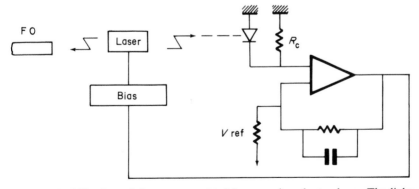

Fig. 8.8. Stabilization of the power emitted by a semiconductor laser. The light emitted from the rear face of the laser is detected by a photodiode. The voltage produced in the load resistance R_c is compared with a reference voltage. The difference controls the bias of the laser to regulate the emission of light

94

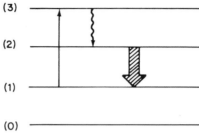

Fig. 8.9. Four-level system. *From left to right*: optical pumping from level (1) to level (3); non-radiating transition from level (3) to level (2) which has a long lifetime; radiating avalanche from level (2) to level (1) at a wavelength greater than that of pumping

states. To obtain this result, the ion to be excited must have at least three available energy levels (see Fig. 8.9).

To raise an electron to level 3, optical pumping can be used in which the necessary quantum of energy is provided by a photon. From level 3 the electron can return to level 1 and re-emit the pumping photon. It can also descend to the intermediate level 2 by a non-radiative transition. This fluorescence level is essential for stimulated emission.

If the duration of the stay in level 2 is sufficiently long, a population inversion occurs after a certain time ($N_2 > N_1$). The population density of level 2 will be greater than that of level 1. It will produce, at a given instant, an 'avalanche' from level 2 to level 1, with emission of light by a relaxation oscillation. This 'avalanche' could equally be caused by stimulated emission by means of a two-mirror Fabry-Pérot cavity with suitable adjustment of the parameters of the system to provide synchronism or continuous operation.

In the Fabry–Pérot cavity one mirror has a reflection coefficient, $R_1 = 100\%$ and the second has a coefficient $R_2 < 100\%$ to allow emission of light to the exterior.

The gain of light amplification, the absorption coefficient and the spectral bandwidth of the ray are connected by expressions which will not be given here.

8.4.2 The YAG–Nd^{+++} laser (Fig. 8.10)

The Nd^{+++} ion in a glass or a crystal such as YAG ($Y_3Al_5O_{12}$ yttrium–aluminium–garnet) provides the emission system. In YAG the Nd ion substitutes for an atom of yttrium and forms a four-level system.

The strongest pumping bands are from 750 to 870 nm and the best laser transitions are from 1050 to 1120 nm and from 1300 to 1400 nm. The optimal concentration of Nd for continuous operation is around 1%.

Continuous operation has been obtained, at normal temperature, at 1064

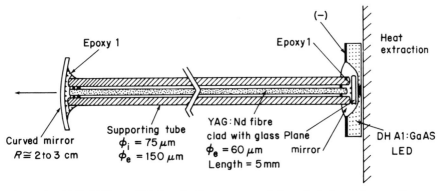

Fig. 8.10. YAG–Nd laser. This longitudinal monomode laser is designed for coupling to a monomode fibre (after J. Stone and C. A. Burrus)

and 1320 nm, providing about 1 mW. The pumping diode is of the Burrus type (see Fig. 8.10).

The rays from the Nd ion have a very narrow bandwidth ($\Delta\lambda_0 < 1$ Å). Nd laser sources are highly reproducible, and the lifetime seems to be very great. They allow operation at a unique frequency and mode and are adaptable to monomode fibres.

8.4.3 Overview

There are other 3 + ions (ytterbium, holmium, erbium and thulium) which can be introduced into crystal or vitreous lattices to provide laser oscillators for optical fibres (Table 8.2).

Table 8.2

Ion	Nd^{3+}	Yb^{3+}	Ho^{3+}	Er^{3+}	Tm^{3+}
$\lambda_0(\mu m)$	1.06, 1.37	1.015, 1.06	2.1	1.536, 1.543	1.85, 2.015

9 Coupling the source to the light guide

9.1 GENERAL

The power available at the output of an optical guide used for communication is small, first, on account of propagation attenuation and second, because of connecting losses and, above all, the input coupling loss between the source and the light guide. This is generally the second in importance in the assessment of the losses of a link, and there are three reasons for this:

(1) The input surface does not intercept some part of the source radiation;
(2) The guide is not closed to rays which enter too obliquely; and
(3) Part of the energy of the rays which are intercepted is reflected from the input face.

All this, in a first approach, is a matter of geometry and refractive index. The most significant characteristics are the limited surface areas and polar diagrams of radiation intensity together with separation and misalignment. Failure of interception causes serious inefficiency, and unfortunately there is little that can be done. The 'optical extent' is conserved, and it is therefore essential to use sources as small as the core of the fibre.

However small sources emit very divergent beams. Consequently the distance between the source and the fibre must not exceed two to four times the diameter of the core. On the other hand, the half-angle of the cone of acceptance generally has a value of 10–14 degrees, even though the half-angle, measured at 3 dB, of the polar radiation diagram varies from 15 to 60 degrees. Several manufacturers produce and market sources already connected to a 'pigtail', a section of fibre to be joined to the light guide by the user.

Loss by Fresnel reflection (which can be exactly calculated) is the smallest, and this can be made virtually negligible by interposing an anti-reflecting layer or a resin to match the refractive indices.

Coupling of an incoherent light-emitting diode and a multimode semi-conductor laser, assumed to be an incoherent source, to a multimode fibre will be examined. Finally, coupling of a laser to a monomode fibre will be summarized.

9.2 PHOTOMETRY AND THE SYSTEM ENERGY

It is accepted that the light sources under discussion have the following two properties:

(1) They are uniform; that is they have the same characteristics at every point on their surface.
(2) They are Lambertian; that is, their radiation obeys Lambert's law (a French mathematician and physicist who established photometry, 1728–77). Hence their luminance (or brilliance) at a point on their surface is the same in all directions. The 'integral radiator' or Plank radiator, also called a 'blackbody', is a Lambertian radiator.

A good example of a luminous source which has the two properties indicated above is the Sun, whose photospheric surface radiates very much as an integral radiator. The photosphere is equally luminous for very oblique emission as for normal emission. Consequently, the Sun appears as a uniformly brilliant disc with scarcely any shadow.

9.2.1 The energy system of global radiation

The energy radiated per second by an element of area dA_s is called the 'element flux energy' $d\phi_e$ in watts (W).

The reader will recall that to measure the solid angle within a cone the area of the portion of a spherical surface cut out by this cone on a sphere of unit radius centred at its apex is evaluated. The unit is the *steradian* (sr)(see Fig. 9.1). Around a point, the maximum solid angle is 4π sr. A plane passing through a point separates two semi-spaces each having a solid angle of 2π sr.

If the cone, which limits the solid angle, is one of revolution with a semi-angle

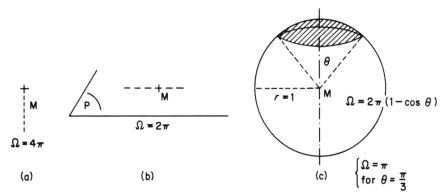

Fig. 9.1. Solid angle (space angle) Ω. This is measured by the surface area which it intersects on the sphere of unit radius centred at its apex. When the directions of the beam are limited by a cone of revolution it can be calculated as a function of the planar angle at the apex of the cone (a) 2π, (b) π and (c) 2θ, respectively. The unit is the steradian, which is dimensionless

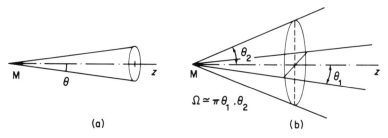

Fig. 9.2. Solid angle. (a) Narrow beam formed by a cone of revolution; (b) elliptic conical beam

at the apex of θ, the solid angle has a value:

$$\Omega = 2\pi(1 - \cos \theta) \tag{9.1}$$

(If Ω has a value of 1 steradian, $\theta \simeq 32.77$ degrees. If Ω has a value of π sr, $\theta = 60$ degrees). If θ is small $\Omega \simeq \pi\theta^2$ (see Fig. 9.2). (9.2) A solid angle may also be called a 'space' angle.

Consider photometric quantities. The energy intensity I_e of a source in one direction within the solid angle $\Delta\Omega$ is the limit of the ratio $\Delta\phi/\Delta\Omega$ of the flux emitted by the source in the solid angle $\Delta\Omega$ and the solid angle itself, as its value tends to zero:

$$I_e \sim \frac{\mathrm{d}\phi_e}{\mathrm{d}\Omega} \quad \text{(in watts per steradian: W sr}^{-1}) \tag{9.3}$$

For a Lambertian source (like the Sun):

$$I_e(\alpha) = I_e(0)\cos \alpha \quad \text{(Lambert's law)} \tag{9.4}$$

Assume now that the propagation medium is homogeneous, non-absorbant, non-emissive and non-diffusing. Under these conditions, luminous energy propagates in a straight line conserving its energy intensity I_e in each direction from the point of emission M (Fig. 9.3).

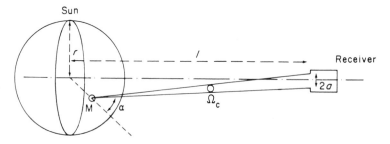

Fig. 9.3. Spherical source of radius r, blackbody radiation. A surface element of area $\mathrm{d}A$, around the point M, emits a beam around the direction α in the solid angle Ω_c which transmits energy flux $\mathrm{d}\phi$ towards the entrance of a receiver at distance l

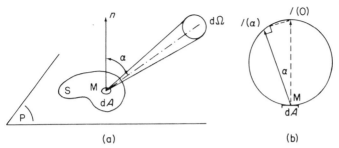

Fig. 9.4. (a) On the source S, the element of area dA, around the point M, radiates a flux element dϕ in the solid angle dΩ which includes the direction forming an angle α with the normal at M. The energy intensity is $I(\alpha) = \mathrm{d}\phi/\mathrm{d}\Omega$ in watts per steradian. If the source is 'Lambertian' $I(\alpha) = I(0)\cos\alpha$, (b) Polar diagram of radiation intensity in a meridional plane

The illuminating energy E_e at a point on a receiving surface is the limiting value of the ratio of the energy flux received by a surface element surrounding this point and the area of the element as its value tends to zero (Fig. 9.4):

$$E_e \sim \frac{\mathrm{d}\phi_e}{\mathrm{d}A_c} \quad \text{(in watts per square metre: W m}^{-2}\text{)} \tag{9.5}$$

It remains to relate the energy intensity of the apparent area of the source d$A_s\cos\alpha$, which is the area of the orthogonal projection of the element of the emissive surface on to a plane perpendicular to the direction under consideration.

The luminance energy L_e is the ratio of the elemental energy intensity and the apparent elemental area around M:

$$L_e(\alpha) = \frac{\mathrm{d}I_e(\alpha)}{\mathrm{d}A_s\cos\alpha} \quad \text{(in watts per square metre per steradian: W m}^{-2}\text{ sr}^{-1}\text{)}$$

$$\tag{9.6}$$

Hence

$$L_e(\alpha) = \frac{\mathrm{d}^2\phi_e}{\mathrm{d}A_s\cos\alpha\,\mathrm{d}\Omega} \tag{9.7}$$

The luminance energy is constant for a Lambertian source, like the Sun:

$$L_e(\alpha) = L_e(0) \tag{9.8}$$

Finally, a last quantity relating to the source is defined: the emittance $M_e = \mathrm{d}\phi/\mathrm{d}A_s$ in watts per square metre (W m^{-2}) at a point M of the source. The emittance has the same value at all points of a uniform source.

It can be shown that, for a Lambertian source, emittance and luminance are related by:

$$M_e = \pi L_e \tag{9.9}$$

Effectively, for a Lambertian radiator:

$$I(\alpha) = \left(\frac{d\phi}{d\Omega}\right)_\alpha = I(0)\cos\alpha$$

It can be seen that:

$$\phi = \int_0^{\pi/2} I(0)\cos\alpha\, \frac{d\Omega}{d\alpha}\, d\alpha \quad \text{with} \quad \Omega = 2\pi(1 - \cos\alpha)$$

It is found that:

$$\phi = \pi \cdot I(0) \tag{9.10}$$

The energy radiated by a source in the solid angle 2π with variable intensity $I(0)\cos\alpha$ is equal to the energy which would be radiated in the solid angle π with intensity $I(0)$ in the normal direction.

The emittance M_e of a blackbody radiator is proportional to the fourth power of the thermodynamic temperature:

$$M_e \sim \sigma \cdot T^4 \quad \text{(Stefan's law)} \tag{9.11}$$

with:

$$\sigma = 5.67 \cdot 10^{-8}\,\mathrm{W\,m^{-2}\,K^{-4}}$$

9.2.2 Energy units for monochromatic radiation

From the magnitude of global radiation G, a monochromatic radiant energy can be defined by:

$$G_\lambda = \frac{dG}{d\lambda} \tag{9.12}$$

The same terms are used with the addition of the adjective 'monochromatic'. In the dimensional equation an additional term L^{-1} is included and m^{-1} for the units.

Hence, the monochromatic energy flux is:

$$\phi_\lambda = \frac{d\phi}{d\lambda}\,(\mathrm{W\,m^{-1}}) \tag{9.13}$$

and the monochromatic luminance:

$$L_\lambda = \frac{dL}{d\lambda}\,(\mathrm{W\,m^{-3}\,sr^{-1}}) \tag{9.14}$$

or, if λ is the only length variable, L_λ can be expressed in: $\mathrm{W\,m^{-2}\,sr^{-1}\,\mu^{-1}}$ or $\mathrm{W\,m^{-2}\,sr^{-1}\,\mathring{A}^{-1}}$.

9.2.3 The solar radiator

Assume that an observer at a distance l from the Sun, in the vicinity of the Earth, has a light receiver with a circular aperture of diameter $2a$ perpendicular to the direction of the Sun. Let L_e be the luminance energy of the photosphere and ΔA_s a small area around a point M of the photosphere, where the capture area forms an angle α with the normal. The received energy flux is:

$$\Delta \phi_c = L_e \cdot \Delta A_s \cos \alpha \cdot \Omega_c \qquad (9.15)$$

where Ω_c is the solid angle of the receiving aperture as seen by the Sun.

If r is the radius of the Sun, the total received flux is:

$$\phi_c = L_e \cdot \pi r^2 \cdot \Omega_c \quad \text{with} \quad \Omega_c \simeq \pi \frac{a^2}{l^2} \qquad (9.16)$$

and the illumination of the receiver is:

$$E_c = \frac{\phi_c}{\pi a^2} = L_e \pi \frac{r^2}{l^2} \qquad (9.17)$$

From equation (9.9) the emittance of the solar source:

$$M_s = \pi L_e \qquad (9.18)$$

Consequently,

$$\frac{M_s}{E_c} = \frac{l^2}{r^2} \qquad (9.19)$$

The photospheric surface behaves like a wave surface surrounding a point source at the centre of the Sun for the problem under consideration and for the condition $l \gg r$.

If l and r are known and E_c is measured, M_s can be determined. Measurement from outside the terrestrial atmosphere gives the following result for a capture area of $1\,\text{cm}^2$:

$$\phi = 2\,\text{cal/min}$$

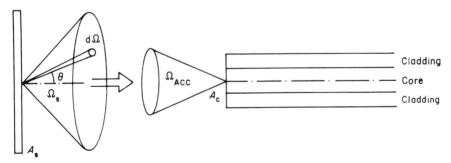

Fig. 9.5. Coupling of a source to a fibre. The source radiates the same energy intensity in all directions within the solid angle Ω_s at all points on the disc of area A_s. The fibre ensures guidance for directions within the solid angle Ω_{ACC} at all points on the cross-section of the core of area A_c for a step index fibre. It is clear that small sources radiating in a narrow angle are preferred. In this way, the efficiency of coupling of the source beam can be maximized

The illumination outside the terrestrial atmosphere is therefore:

$$0.14 \, \text{W/cm}^2$$

The emittance of the solar photosphere is:

$$6.45 \, \text{kW/cm}^2, \text{ for the whole spectrum}$$

The attenuation l^2/r^2 has a value of 46.6 dB.

These magnitudes, which relate to our environment, are interesting, at least for comparison purposes, with quantities which will be discussed in connection with fibre technology, particularly energy fluxes per unit area (illumination, emittance).

Notice that bad matching of a source to a fibre can lose up to 30 dB (Fig. 9.5).

9.3 THE LIGHT-EMITTING DIODE (LED) IN NORMAL RADIATION

This is the simplest light source to connect. It presents a plane uniformly illuminated circular disc whose luminance does not depend on the observation angle. In other words, this source can be regarded as a Lambertian radiator in air for all solid angles $\Omega_s \leqslant 2\pi$.

The diode and the optical fibre together form a symmetrical cylinder of revolution, and they are arranged to have the same axis $0z$ by directing the light emission to the input cross-section of the fibre.

It remains to examine the coupling conditions (see Fig. 9.6). The elemental energy flux radiated by the source in the solid angle $d\Omega$ in the θ direction is:

$$d\phi = L_e A_s \cos \theta \, d\Omega \tag{9.20}$$

where A_s is the area of the luminous source. Hence:

$$d\phi = L_e A_s 2\pi \sin \theta \cos \theta \, d\theta \tag{9.21}$$

If the LED is applied to the core of the fibre and has a diameter less than, or at most equal to, the diameter $2a$ of the core of the fibre, the power entering the cone of acceptance of the step index fibre will be:

$$\phi_{\text{fibre}} = 2\pi L_e \cdot A_s \int_0^{\theta_{\text{ACC}}} \sin \theta \cos \theta \, d\theta \tag{9.22}$$

(One can imagine an infinitely thin lamina of air separating the source and the fibre.)

The power coupled to the core of the SI fibre is:

$$\phi_{\text{fibre SI}} = \pi L_e \cdot A_s \sin^2 \theta_{\text{ACC}} \tag{9.23}$$

$$\phi_{\text{fibre SI}} = \underbrace{2\pi L_e A_s}_{a} \underbrace{n_1^2 \Delta}_{b} \tag{9.24}$$

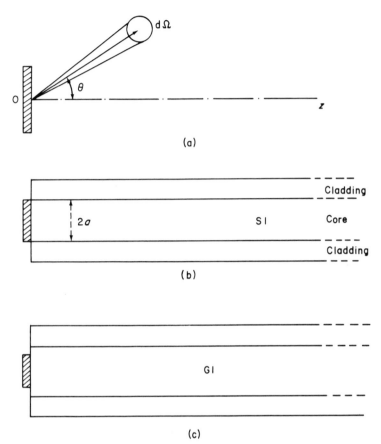

Fig. 9.6. (a) A diode, a Lambertian radiator, of area A_s and luminance L_e emits flux $d\phi = L_e.\,A_s.\cos\theta.\,d\Omega$ in the element of solid angle $d\Omega$; (b) the diode is applied directly to the core of the fibre; for a step index its useful diameter can reach that of the core $2a$; (c) for a graded index, the coupling coefficient is improved if the diameter of the diode is less than $2a$. Centering is critical

(1) $2\pi L_e A_s$ is the total luminous power of an ideal LED radiating into a hemisphere (recall the condition $A_c \geqslant A_s$);

(2) $n_1^2\Delta$ is the coefficient of coupling which depends only on the refractive indices of the guide. It is equal to $1/2(NA)^2$.

If $\Delta = 1\%$ and $n_1 = 1.48$ the coupling coefficient is 0.022. The total luminous power loss of the LED is 16.6 dB. For a graded index fibre, with a parabolic law $g = 2$ and for an LED having the same diameter $2a$ as the core of the fibre, the coupling coefficient is $(1/2)n_1^2\Delta$.

It is important to emphasize that the alignment of the LED along the axis of the core of the fibre (SI or GI) is critical; longitudinal displacement is less so.

9.4 THE LIGHT-EMITTING DIODE (LED) WITH LATERAL RADIATION AND THE LASER DIODE

Far-field measurements on the emission of light-emitting diodes with radiation normal to the plane of the $p–n$ junction clearly show an intensity distribution diagram close to that of cos α. However, more directional radiation is obtained, following $(\cos \alpha)^x$, with $x > 1$, for double heterojunctions with lateral radiation. More directional emissions are obtained with laser diodes, again with lateral emission.

The width and length of the emitted bright spots have a maximum difference in the orthogonal planes $z0x$ and $z0y$ containing the principal direction of emission $0z$. Source dimensions are small and the beams are narrow, which favours introduction of light into the guide even if it is an optical fibre.

The distorted beam can, in principle, be transformed into an approximately symmetrical one by a cylindrical lens (or better, by a compact set of graded index prisms). Having achieved this, the light source appears as a small brilliant surface of area A_s radiating an approximately uniform intensity in the solid angle Ω_s, limited to 3 dB (Fig. 9.5). Hence the beam of rays emitted by the source has an extent S:

$$S_{source} = A_s \cdot \Omega_s \quad \text{(in air)} \qquad (9.25)$$

The fibre presents a core surface of area A_c which accepts all rays within a solid angle Ω_{ACC}, in air, or Ω_c in the core at each point, for a step index fibre. Hence the beam of rays accepted by the core has an extent:

$$S_{fibre} = A_c \cdot \Omega_{ACC} \quad \text{(in air)} \qquad (9.26)$$

In these conditions, if the luminance of the source is L_e, the maximum flux which can be directly introduced into the fibre is:

$$\phi_c = L_e A_{min} \cdot \Omega_{min} \qquad (9.27)$$

A_{min} is the smaller of the two areas A_s and A_c which are applied to each other and Ω_{min} is the smaller of the two solid angles Ω_s and Ω_{ACC}.

If the quantities A_{min} and Ω_{min} relate to the same component, source or fibre, no improvement is possible. In effect, geometrical optics can only conserve the extent $n^2 \cdot S$. In these conditions, either $\phi_c = L_e \cdot S_{source}$ or $\phi_c = L_e \cdot S_{fibre}$.

In the case where the quantities A_{min} and Ω_{min} do not relate to the same component, matching by lens is possible, so that S_{min} becomes equal to S_{source} or S_{fibre}, which remain constant.

*$n^2 S$ is the optical extent such that S is the geometrical extent.

9.5 COUPLING A LASER INTO A MONOMODE FIBRE

Electromagnetism leads to consideration that the em field in the core of the fibre is inseparable from the field in the cladding and that the em wave of a guided mode is inseparable from the evanescent wave which accompanies it in

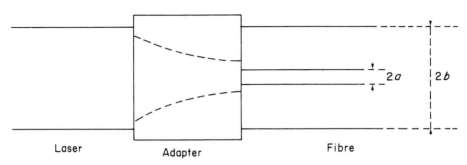

Fig. 9.7. Diagram of the principle of excitation of the fundamental mode of a step index fibre by a Gaussian laser beam. In general the adaptor is a cylinder whose length is of the order of 1000λ or a cylindrical lens. The ratio of input and output diameters does not exceed 7

the cladding. In a unique fundamental mode the energy propagated in the cladding, whose diameter is around ten times that of the core, is important.

The output beam of an injection laser does not have cylindrical symmetry. This distorted beam can, in principle, be transformed to a beam of symmetrical Gaussian profile by a system of cylindrical lenses (Fig. 9.7).

A monomode solid state laser can emit an approximately Gaussian beam:

$$I(r) = I(0) \exp\left[-\left(\frac{r}{w}\right)^2 \right] \qquad (9.28)$$

Calculation and experience show that such beams can be aligned along the axis of a monomode fibre and matched to the cut of its fundamental mode, and the coupling coefficient can be high. However, in practice it remains difficult to obtain efficiencies of 90–95%.

10 Optical detectors (for fibre communication)

10.1 GENERAL

As already mentioned, photon–material interactions are complex and of many kinds. They depend principally on the wavelength or, more explicitly perhaps, the energy of the photon.

The optical detector, in general, plays a crucial role in optical communication systems since it must satisfy the requirements of compatibility very closely. In systems which use optical fibres the light detector is a component just as essential as the source. The principal requirements concern sensitivity, bandwidth and the noise which is introduced. The space required, coupling to the fibre (or bundle of fibres), associated electronic circuits, power requirements and price are also to be considered.

Among detectors which can be used in optical fibre systems are:

(1) *The vacuum photodiode*: This uses the photoelectric effect. The photocathode emits electrons only if the photons have sufficient energy. One part of the energy $h\nu$ is needed for the electron to reach the surface of the cathode, another is necessary for it to leave the surface and the rest is carried by the electron in kinetic form. The limit of efficiency occurs around $1.2\,\mu m$ with complex materials, giving quantum efficiencies around 20%.

(2) *Photomultipliers*: When photocathodes are made from III–V compounds, attractive sensitivities are obtained. Their size, high-voltage requirements and price are disadvantages.

(3) *Volume effect semiconductor diodes*: This uses the inverse process to that of light emission in LEDs and SCLs. As with the emitters, these receiving diodes have a great advantage—they are very small.

As long as the energy difference between the conduction and valence bands of a diode is fixed, to a first approximation, the energy of the photon emitted by a forward-biased diode is the same as the minimum quantum energy $h\nu$ of a photon captured by the same diode in reverse bias.

The cutoff wavelengths λ_c, the upper boundary of the usable spectrum is

106

Table 10.1 Band gaps ΔE and cutoff wavelengths for different elements and compounds

In group IV					Binary compounds		
ΔE(eV)	$\lambda_c(\mu m)$					ΔE(eV)	$\lambda_c(\mu m)$
C	7	0.18	(II, VI)	Zn	S	3.6	0.34
Si	1.1	1.13		Al	Sb	1.52	0.81
Ge	0.72	1.72	(III, V)	Ga	As	1.43	0.87
Sn	0.08	15.5		In	P	1.25	0.99
				In	Sb	0.2	6.2

Ternary compounds $Ga_xIn_{1-x}As$	Quaternary compounds $Ga_xIn_{1-x}As_yP_{1-y}$
ΔE: $1.43 - 0.36$ (eV) λ_c: $0.87 - 3.44(\mu m)$	ΔE: $1.36 - 0.36$ (eV) λ_c: $0.92 - 3.44(\mu m)$

(Table 10.1):

$$\lambda_c \simeq \frac{1.24}{\Delta E} \quad (\lambda_c \text{ in } \mu m \text{ and } \Delta E \text{ in eV}) \tag{10.1}$$

In the great majority of cases at present reception of luminous signals transmitted by optical fibres is direct and aperiodic. In fact, coherent detection, heterodyne or homodyne, requires that the phase of a pure sinusoidal signal is maintained as a reference. Such detection is of great interest in open optical communication through space in the presence of light from the sky.

If the luminous signal is propagated in a closed guide interest is less because the composition of the fields, in amplitude and phase, propagated by numerous modes with different delays makes phase comparison useless. Coherent equipment is limited in practice to monomode fibres and ribbons.

Alternatively, using Heisenberg's uncertainty principle it has been shown that the lower limit of photon noise, referred to the input of an optical amplifier, can be expressed in terms of an equivalent temperature by the approximate relation:

$$T_{min} \simeq 1.4 \frac{h\nu}{k} \tag{10.2}$$

when the gain of the amplifier is greater than 3 or 4:

$$\lambda_c = 1\,\mu m, \quad \nu = 300\,THz \quad \text{and} \quad T_{min} = 20\,000\,^\circ K$$

The optical amplifier is therefore subject to considerable noise! Furthermore, unlike amplifiers, the performance of detectors increases with frequency. Consequently, in optics, detection is direct, without preamplification, in contrast to normal radio practice.

Finally, the present choice lies with volume effect semiconductor diodes. The apparently major disadvantage of these photodiodes is that their bandwidth is often limited to several hundreds of MHz, while future systems must operate at rates of several Gbit/s.

10.2 OPERATION OF PHOTODIODES

10.2.1 Absorption of light

A reverse-biased photodiode, of whatever type (PIN, avalanche, etc.) contains a depletion region (DR) of free electric carriers subjected to a high electric field ($\simeq 10^4$ V/cm) between two regions of p and n, respectively, where the electric charge is zero and the electric field is very weak.

Photons of higher energy than the width of the forbidden band of the semiconductor are absorbed by the creation of an electron–hole pair:

(1) Every pair created in the DR is separated by the field and joins the current by the flow of an electron in the external electric circuit.
(2) An electron–hole created in the p or n region is a minority carrier which diffuses as far as the DR, drawn by the field, unless it recombines with a hole or majority electron. In this case, the photon will not have contributed usefully to the current.

10.2.2 Quantum efficiency

Photons absorbed in the DR have a quantum efficiency $\eta = 1$. Photons absorbed outside the DR can participate in current only if they are absorbed at less than a diffusion length from the limits of the DR.

In total, the quantum efficiency

$$\eta = \frac{\text{Number of pairs}}{\text{Number of photons}}$$

is less than 1. Three factors reduce the quantum efficiency; (1) losses by reflection at the input interface, (2) recombination in the volume of created carriers and (3) recombination on the surface which creates leakage currents and which can be effectively prevented by means of guard rings. Reflection losses are equally countered by means of a supplementary layer called 'anti-reflecting', which can be optimized to suppress all reflection in normal incidence at a given wavelength, (Fig. 10.1).

There remains the risk of the production of pairs and their recombination without taking part in the current. These detectors are quadratic since they respond to the intensity of the light averaged over a large number of periods. The detected current varies linearly with the incident optical power.

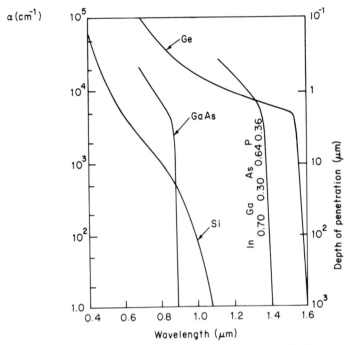

Fig. 10.1. Absorption coefficient of light α (cm^{-1}). At the depth of penetration, $1/\alpha$, the normally incident luminous flux, measured immediately below the input interface, has lost 63% of its value, that is, it has been divided by e. The curves correspond to the materials most commonly used in photodetectors

10.2.3 The speed of response

For absorption in the DR speed is limited by the transit time of the carriers in this region:

$$t_t = \frac{e}{v_{lm}}$$

where e is the useful thickness of the DR and v_{lm} is the limit of the drift velocity (about 10^7 cm s^{-1}).

For absorption outside the DR it is necessary to take account of the diffusion times of the minority carriers up to the DR:

$$t_{\text{diff}} = \frac{e}{v_{\text{diff}}} = \frac{1}{\alpha^2 \cdot D} \quad (D \text{ is Einstein's constant})$$

The appearance of a dipole $+Q-Q$ has an equivalent effect to that of a capacitance C which varies with the bias voltage and the useful size e of the DR. The best compromise must be found between these different effects. For a bandwidth of 3 GHz, e is of the order of $10\,\mu m$.

10.2.4 Sensitivity (S_e)

The global efficiency of conversion of luminous power is characterized by an electric current. The general expression for the conversion of an incident flux of photons of the same quantum hv into a primary photocurrent I_ϕ is given by:

$$I_\phi = \eta \cdot q \frac{P_0}{hv} \qquad (10.3)$$

where η is the quantum efficiency, q the charge on the electron and P_{opt} the incident optical power.

The sensitivity is the ratio:

$$S_e = \frac{I_\phi}{P_0} \qquad (10.4)$$

of the primary photocurrent and the incident optical power. Consequently:

$$S_e = \frac{q}{hv} \eta \quad S_e(\text{ampere/watt}) = 0.80 \cdot \eta \cdot \lambda_0(\mu m) \qquad (10.5)$$

At a given frequency, S_e is directly proportional to η. In order to optimize the bandwidth it is found that e approaches $1/\alpha$, which gives a quantum efficiency around 60% and, for a wavelength $\lambda_0 = 1\,\mu m$, a sensitivity $S_e = 0.5\,\text{A/W}$.

10.3 THE PIN PHOTODIODE

In this photodiode the compromise between the values of the characteristic parameters leads to a reduced doping of the N region in order to increase the dimensions of the region of space charge. In the limit, an intrinsic material I is achieved with which an N region of low resistivity is associated.

The PIN structure is very widely known. Its sensitivity is very high and its efficiency is also high. In reception it operates as a current generator. To produce a photoreceiver it is necessary to make a current–voltage conversion by means of a high-value load resistance followed by amplification of the voltage.

In this way, the input circuit has a long time constant $\tau = R_1 C_1$ and a small thermal noise (Figs 10.2 and 10.3):

$$\langle i_{th}^2 \rangle = \frac{4kT}{R_1} B \qquad (10.6)$$

Association with a field effect transistor allows the detection of very small luminous powers (some tens of picowatts). Moreover, it is possible to compensate for the fall in the response at high frequencies:

(1) By an inverse filter or equalizer; in short, a differentiating circuit rC with short time constant placed at the output of an amplifier described as 'integrating'; or

(2) By negative feedback, by means of a resistance R_c, in order to increase the

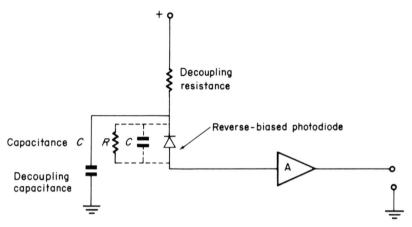

Fig. 10.2. Input circuit for a photodiode. The internal resistance, R, of the order of 500 MΩ, and capacitance, C, of the order of 5 pF, of the photodiode are shown by the broken line

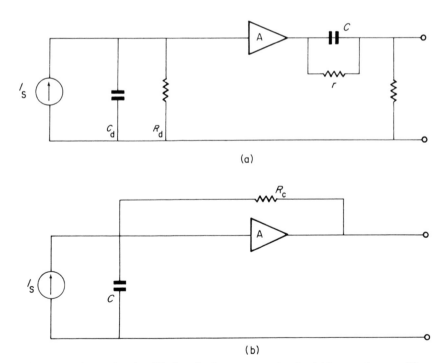

(a)

(b)

Fig. 10.3. Alternative simplified equivalent current circuits. (a) Integrating amplifier; (b) transconductance amplifier

112

pass band at the expense of gain, for a constant gain–bandwidth product. An amplifier of very high gain and small input current is called 'trans-conductance'.

In summary, the PIN photodiode has a very simple structure and good dynamic linearity from several picowatts to tens of milliwatts. It is simple to use and moderately priced. However, progress with photodiodes which have an internal gain, and are called 'avalanche', could favour use of the latter in wideband systems which justify the complication that they impose.

10.4 THE AVALANCHE PHOTODIODE (APD)

When the electric field is sufficiently high (greater than 10^5 V/cm) the carriers drawn into the depletion region (DR) by the field can acquire sufficient kinetic energy to cause ionization by impact and thus create new carriers which are themselves accelerated by the field.

There is therefore an amplification for which the multiplication factor M depends on the magnitude of the electric field. The gain–bandwidth product of this process is limited by the transit time in the amplification region.

However, random fluctuations of the coefficient M about its mean value constitute an additional source of noise. Account is taken of this increase of noise, introduced by the internal gain process by means of a factor F(M) which is of the form:

$$F(M) = M^x \tag{10.7}$$

The exponent x depends on the ratio of the ionization coefficients of electrons and holes then on the material (Fig. 10.4).

The avalanche voltage is approximately proportional to the depth of the depletion region. The P π PN photodiode is a particular case of avalanche

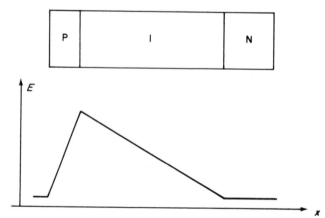

Fig. 10.4. Electric field E in the different regions of a PIN diode

photodiode, which attempts to combine the advantages of PIN and avalanche photodiodes.

10.5 NOISE IN PHOTODIODES

An understanding of the origin, the characteristics and the effects of different kinds of noise is essential for the evaluation and specification of the operational characteristics of the overall optical communication system.

10.5.1 Types of noise

Figure 10.5 shows the different forms of noise which arise in the course of detection and amplification of the signal. Background noise, which is important in open propagation, is negligible with guided propagation if guidance starts in the immediate proximity of the source and if the guide is impervious to external radiation. Beat noise is produced in the detector by the different spectral components of the carrier (the broad line of an LED, for example). Quantum noise, dark current noise and surface leakage current noise are all shot noise and have Poissonian statistics.

Quantum noise, which arises from the intrinsic fluctuations in the optical

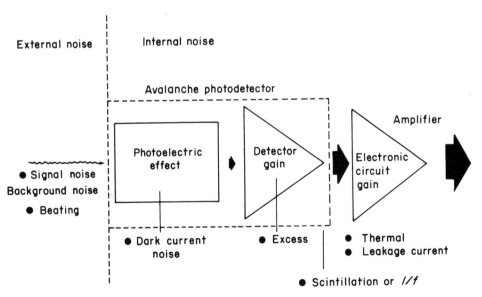

Fig. 10.5. Diagram of the different sources of noise which appear in the course of optical detection and amplification. The intrinsic noise power present at the output in the absence of a signal is called the 'dark' noise. The minimum exploitable signal power is defined as being equal to the dark noise power and it is called the 'equivalent noise power' (ENP) $= W_n/\sqrt{(\Delta f)}$. (*Note*: signal, background and dark current noise are 'shot noise')

production of charge carrier pairs, is fundamental. In a photodiode without avalanche gain, thermal noise, which appears in the load resistance of the detector and in the active elements of the electronic amplifier, is dominant.

Avalanche gain in a photodiode, which is a random process, introduces an additional noise, called 'excess' noise, in the receiver. This noise increases the primary shot noise in proportion to the avalanche multiplication. In spite of this, avalanche diodes provide a noticeable improvement in optical reception by direct detection.

10.5.2 Relative importance and effect of various forms of noise

Figure 10.6(a) shows the equivalent circuit of the front end of an optical receiver; the reduced circuit at (b) shows the signal source i_ϕ and the principal sources of noise.

Let:

$p(t)$ be the instantaneous optical power incident on the photodetector,

C_d the capacitance
R_1 the load or bias resistance $\Big\}$ of the detector

C_a the capacitance
R_a the resistance $\Big\}$ at the input of the amplifier which follows the dectector

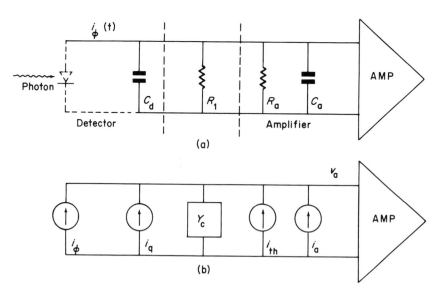

Fig. 10.6. Illustration of the relative importance of the effect of different noise sources. (a) Equivalent circuit of the front end of an optical receiver; (b) reduced circuit, showing the signal source and the principal sources of noise—i_q quantum, i_{th} thermal, i_a amplifier

$i_\phi(t)$ the instantaneous primary photocurrent in the detector:

$$i_\phi(t) = \eta \frac{q}{h\nu} p(t) = S_e \cdot p(t)$$

S_e the sensitivity:

$$S_e = \frac{\eta q}{h\nu}$$

I_ϕ the primary d.c. component, the mean of $i_\phi(t)$,
i_ϕ the primary component of the signal $i_\phi(t) = I_\phi + i_\phi$,
m the avalanche multiplication factor, which varies, and
$M = \langle m \rangle$ the mean of the set of values of m,

The d.c. primary current I_ϕ produces quantum noise whose quadratic mean, after the avalanche gain, is:

$$\langle i_q^2 \rangle = 2q I_\phi \langle m^2 \rangle \cdot \Delta f$$

where $\langle m^2 \rangle$ is the quadratic mean of the internal gain and Δf is the effective bandwidth of the noise.

The quantum noise due the dark current I_n, multiplied by the gain, behaves in the same way and can be combined with the current I_ϕ, to give the total shot noise, which is:

$$\langle i_q^2 \rangle = 2q(I_\phi + I_n)\langle m^2 \rangle \cdot \Delta f$$
$$= 2q(I_\phi + I_n)M^2 \cdot F(M) \cdot \Delta f$$

The index n is relative to the dark current.

(*Note*: a bar above a symbol denotes the mean with respect to time and brackets $\langle \rangle$ the mean of a set of values.)

$F(M)$ is a measure of the degradation due to avalanche multiplication compared with noise-free multiplication.

Thermal noise created in the load or bias resistance R_1 is given by:

$$\langle i_{th}^2 \rangle = \frac{4kT}{R_d} \Delta f \tag{10.7}$$

where k is Boltzmann's constant and T is the absolute temperature.

Sources of noise associated with the active elements of the amplifier can be represented as a series of noise voltages and a current $\langle i_a^2 \rangle$ in parallel. This last term includes the thermal noise associated with the input resistance R_a of the amplifier.

In Figure 10.6(b) the admittance Y_c represents the combined shunt resistances and capacitances of Figure 10.6(a). At this stage, the following amplifier is assumed to be perfectly noise-free.

116

Finally, referring to Figure 10.6(b), the signal-to-noise ratio at the input of a noise-free amplifier can be written in the form:

$$\left(\frac{S}{N}\right)^2 = \frac{i_\phi^2 M^2}{2q(I_\phi + I_n)M^2 \cdot F(M)\cdot\Delta f + (4kT\cdot\Delta f/R_d) + \langle i_A^2\rangle} \tag{10.8}$$

where $\langle i_A^2\rangle$ is the noise associated with the amplifier:

$$\langle i_A^2\rangle = \frac{1}{\Delta f}\int_0^{\Delta f} (\langle i_a^2\rangle + \langle v_a^2\rangle|Y_c|^2)\mathrm{d}f \tag{10.9}$$

Here, f is the frequency and $\langle i_a^2\rangle$ and $\langle v_a^2\rangle$ are assumed to be uncorrelated.

A value of M exists which maximizes the signal-to-noise ratio. For silicon avalanche photodiodes, the optimum value of M is 2.5 for $B \simeq$ several MHz and 100 for $B \simeq$ several GHz. For a photodiode without avalanche gain, $M = F(M) = 1$.

10.6 SPEED OF RESPONSE AND TRANSIT TIME

The speed of response of a photodiode is ultimately limited by the time taken by the charge carriers to cross the depletion region. The electric field in this region is normally greater than 2.10^4 V/cm so the carriers attain a velocity greater than 10^7 cm/s (in the silicon). For a distance of $10\,\mu m$ the transit time is less than 0.1 ns.

10.7 THE STATE OF THE ART IN PHOTODIODE MATERIALS

In the band of wavelengths which extends from the visible to around $1\,\mu m$ the preferred material is silicon. PIN and avalanche photodiodes which give a good performance with AlGaAs sources at $0.85\,\mu m$ are available on the commercial market.

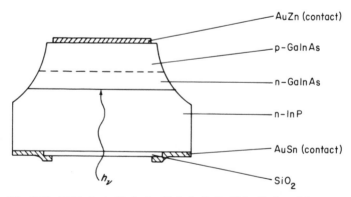

Fig. 10.7. PIN photodiode in GaInAs/InP. This diode, of inverse mesa type, has a diameter of 150 μm. It is fabricated by epitaxy in the liquid phase and by a photolithographic process, and it covers the 960–1600 nm band. The lower layer of SiO₂ is anti-reflectant

Germanium is the preferred candidate for wavelengths greater than $1\,\mu m$. However, the avalanche multiplication excess noise is relatively high, with $F = M/2$.

Photodiodes have been produced with GaAlSb/GaSb heterojunctions in a binary–ternary system covering the waveband from 1.1 to $1.5\,\mu m$. The quantum efficiency is greater than 50% and the avalanche gain is 10–15 for a rise time of 60 ps. GaInAs/InP pin diodes have also been made which operate with a bias of $-3\,V$ at $1.20\,\mu m$ with a quantum efficiency of 60% (Fig. 10.7).

10.8 PERFORMANCE OF PHOTODIODES IN RECEIVERS

The performance of a photodiode in an optical communication system is determined not only by its own characteristics but also by those of the following

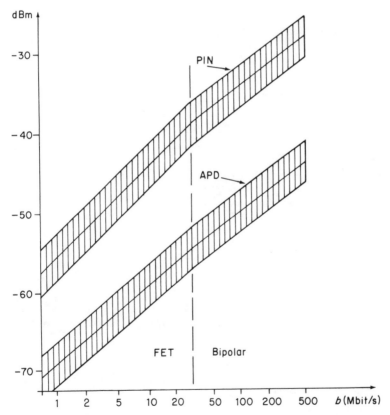

Fig. 10.8. The mean optical power in dB above the mW necessary for reception in order to obtain an error rate of 10^{-8} as a function of the digital data rate in Mbit/s for PIN diodes and avalanche diodes associated with a field effect transistor (FET) and a bipolar transistor. These theoretical and practical results, obtained for silicon photodetectors at $0.84\,\mu m$ and for germanium at $1.30\,\mu m$ are very close to experimentally measured values[4,5]

amplifier. General results are presented in Fig. 10.8 with binary rate b in bit/s as abscissae and the mean power produced in the detector by the optical signal as ordinates, to obtain an error probability of 10^{-8} at the receiver.

The two bands represented show results calculated from the current parameters of silicon field effect transistors (FET) and bipolar transistors which are used in preamplifiers immediately after the detector. The upper band relates to PIN photodiodes and the band below to optimum gain avalanche photodiodes, from 1 Mb/s to 500 Mb/s.

Agreement between theory and practice is satisfactory. At $1.3\,\mu m$ the excess noise of the germanium avalanche photodiode degrades the performance of this component by about 6 dB in comparison with that of the silicon avalanche photodiode.

Limitation of receiver sensitivity by quantum noise inherent in the signal is represented by a mean number of 11 photoelectrons per bit.

11 Modulation, light modulators and switches

11.1 GENERAL

In order to transmit intelligible information over any communication channel it is necessary to cause a change in the operation of the channel. This change is called 'modulation', and it is modulation which transmits the 'message'. For example, if shepherds on the hills wish to alert the villagers down in the valley without using sound waves they could close the valve of the pipe which feeds water to the village fountain; this is a kind of modulation which transmit 'coded information' which, when decoded, is the 'message'.

If the transmission of information requires modification or modulation of an operating characteristic of the communication channel there is an important problem to choose the characteristic which is most favourable or most convenient to modify.

For a continuous or semi-continuous optical channel the characteristic chosen could be amplitude, phase, frequency, polarization or direction of propagation. For pulsed operation of the channel the chosen characteristic could be amplitude, duration or position of the pulse within a time slot.

By rapidly bringing the output of the water channel to 'zero' the shepherds in the above example have used an indeterminate modulation format on the limit between the two possible formats, the 'analogue' and the 'digital'. By regulating the water output to be 'zero' or 'one' during each five minutes of an hour they could send coded information by producing a sequence of binary digits. Hence the format would be digital. By varying the output of the water channel in a continuous manner following the function $d = a + b \sin \omega t \, (a \geqslant b)$ they could use analogue amplitude modulation by transmitting, laboriously, for a period T and varying this period.

The water channel which we have taken as an example in order to explain the terminology seems better suited to a digital rather than analogue format. It will be seen that it is the same for a fibre optic channel.

11.1.1 Characteristics which can be modulated

The characteristics of light waves propagated in an optical fibre appear to be numerous, but few can be modulated.

The optical fibre, monomode or multimode, attenuates the signal and disperses it. Now, it is required to transmit a signal covering a certain minimum band of the spectrum; consequently the optical fibre must allow at least this band to pass. The result is a throughput limit L which is imposed either by the threshold power level acceptable for reception or by the pass band of the fibre of length L.

In general, polarization, if it is determined when the light is injected into the fibre, is not maintained throughout propagation because of mechanical constraints inherent in the fibre and other irregularities such as ellipticity of the core/cladding interface whose cross-section is not precisely circular. These cause coupling of orthogonal polarization modes and, consequently, rotation of polarization. All signal processing which requires a defined and predictable polarization at the receiver is therefore to be avoided.

This problem of the transmission of polarization is important. No doubt it could be resolved for cylindrical fibres over very short distances or, more easily, for optical ribbons. If it were resolved, it would be possible to implement heterodyne detection necessary for phase or frequency demodulation. It would also be necessary to use monochromatic monomode sources of light both for transmission and reception, as a local oscillator.

At present, amplitude modulation and direct asynchronous detection seem to be the best solution for optical fibres.

11.1.2 Choice of format

The two formats, analogue and digital, can be obtained, and the analogue format has the merit of simplicity. However, signal-to-noise ratios limit its use to relatively narrow bandwidths and short distances. The digital format provides an immunity to noise at the expense of bandwidth, which can be convenient. Consequently it is appropriate for transmission by fibre in which wide bandwidths are available. Most applications of fibre optics over medium and long distances require digital pulse modulation.

Amplitude modulation is simple to achieve and suits diode and injection laser sources which can be modulated directly by variation of their control current to achieve rates of Gbit/s.

However, future systems could use other types of light source such as the N_d^{3+} laser. These systems will require the use of external modulators.

11.1.3 External light modulators

External modulators already constitute a group of optical components which are necessary for the full exploitation of light guides. These components will be used as modulators downstream of a continuous source of light, and will serve equally as time-division multiplexers and switches for light guides. They consist essentially of one or several relatively short segments of light guide where the required operation is achieved separately from the long guides used for transmission and within the overall system requirements. In these modulators propagation of light over a short length is modified by the application of an action which has a strong effect. (In contrast, in a measuring instrument, when a small action is to be detected a long length of light guide is used.)

For example, one induces:

(1) A coupling between two guides, to transfer light from one to the other;
(2) A variation $\Delta\beta$ in the wave number of the fundamental mode propagated in the guide, to modify the phase. Along the length L over which the variation of refractive indices has caused an increase $\Delta\beta$, the increase in phase is $\Delta\phi = \Delta\beta \cdot L$. Electro-optic, magneto-optic or acousto-optic effects are used to achieve this. Amplitude modulation is also obtained by causing a phase-modulated wave to interfere with an unmodulated wave of the same amplitude. The combination of these two waves, of the same frequency, has an amplitude which can vary from 0 to 1 (Figs 11.4, 11.5 and 11.7).
(3) Appearance in the propagating medium of a diffraction grating produced by an acousto-optical or electro-optical effect which deviates the light to pass, or not, from guide A to guide B (Figs 11.8 and 11.9).
(4) A displacement in the spectrum of the wavelength which limits the transparency band of the propagating medium by the action of a strong electric field to produce absorption of light in the medium and, by causing it to cease, sudden transparency.

All these components, which are mainly plane pieces of guide, have one great advantage; they are easy to fabricate by photolithographic techniques and are compatible with monomode fibres.

Certain applications will stimulate research and, no doubt, fibres capable of maintaining light in the same state of polarization over long distances and components whose operation is independent of polarization will be developed.

11.2 COUPLING OF TWO LIGHT GUIDES BY EVANESCENT WAVES

In Chapter 4 guidance of light by a cylindrical dielectric rod was studied; it was of unlimited length along the $0z$ axis, of dielectric constant ε_1 and surrounded by an external medium of dielectric constant ε_2 and unlimited length and radius. The two media are non-dissipative, that is, light is not absorbed anywhere.

The question was whether light could propagate without attenuation in such a guidance system, and the answer is affirmative. With the restriction $\varepsilon_1 > \varepsilon_2$, there is always at least one solution. The possibilities are expressed by:

(1) A limited series of progressive modal waves $\psi_m \cdot e^{j(\omega t - \beta_m z)}$;

(2) An equally limited series of regressive waves $\psi_m \cdot e^{j(\omega t + \beta_{\dot{m}} z)}$ which express the reciprocity of the model.

Hence, em energy can propagate longitudinally without attenuation in a perfect guide from $z = -\infty$ to $z = +\infty$, or vice versa. This implies that energy is not radiated transversely. However, the em field exists in the exterior medium and extends to infinity since its amplitude decreases exponentially. Nevertheless, the em energy does not propagate transversely. Such is the result.

Consider an irregularity, called an 'absorbant centre', placed in the exterior medium, not too far away on account of the exponential decrease and occupying a small region of space; it is necessary to add new boundary conditions at the surface of the irregularity to those already imposed for the cylindrical interface which the field must observe. With these new conditions, the field will be modified as much in the rod as in the exterior medium and em energy will propagate towards the irregularity where it will be lost. The wave in the exterior medium will no longer be purely evanescent and the dielectric rod will radiate a small amount of light towards the centre. This energy will be lost from the light propagating along $0z$.

Now consider another physical model consisting not only of the previous guide, which will be called A, but also a second guide B identical to A, deduced from A by a side travel not parallel to $0z$.

We look for the conditions under which the two coupled guides can propagate light without attenuation, taking account of the fact that each guide represents an irregularity to the other which causes a transverse propagation of energy by which the coupling is achieved. An exact solution to this problem is not known.[8]

However, when the two elementary guides are clearly separated, that is, when the cross-section shows interfaces in the form of two non-intersecting circles, a perturbation theory approach may be used which provides two fundamental solutions to the problem of light propagation in the system of two identical parallel guides:

$$\psi_a(x, y)e^{j(\omega t - \beta z)} \quad \text{and} \quad \psi_b(x, y)e^{j(\omega t - \beta z)}$$

are the fundamental modal solutions which satisfy the Helmholtz scalar equation for the isolated guides A and B, respectively, and the fundamental modal solutions relating to the composite guide are:

$$\psi_a(x, y)e^{j(\omega t - \beta z)} \quad \text{and} \quad \psi_b(x, y)e^{j(\omega t - \beta z)}$$

with:

$$\psi_+ = \psi_a + \psi_b \quad \text{and} \quad \psi_- = \psi_a - \psi_b$$
$$\beta_+ = \bar{\beta} + C \quad \text{and} \quad \beta_- = \bar{\beta} - C$$

This result is obtained when the orientations of the electric fields in the cylindrical rods conform to those indicated in Fig. 11.1 for the symmetrical mode $+$ and for the asymmetric mode $-$. In the two cases shown, with a unique fundamental mode for each guide, the electric fields outside the rods will be everywhere approximately parallel to $0x$ on condition that ε_1 is a little

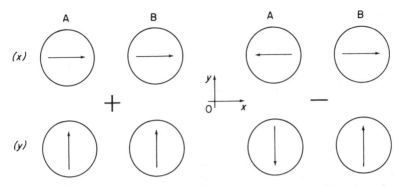

Fig. 11.1. Fundamental propagation modes of two identical dielectric rods. Orientation of the transverse components of the electric field for each of the four modes. For a single isolated rod at a given z and a given r there is practically no variation of the field as a function of azimuth, in the fundamental mode if the guidance is weak

greater than ε_2 (this is the condition for weak guidance). Hence, in spite of the small discontinuity in the normal component of the electric field in crossing the interface, the component and resultant electric fields will be approximately parallel to $0x$.

For parallel polarized modes, the transverse component of the resultant electric field will have an amplitude:

$$E(x, y) = M_+ \cdot \psi_+ e^{j(\omega t - \beta_+ z)} + M_- \cdot \psi_- e^{j(\omega t - \beta_- z)} \tag{11.1}$$

where M_+ and M_- are the amplitudes of the $+$ and $-$ modes.

If for abscissa $z = 0$ in the composite guide, rod A is illuminated with unit power and rod B with zero power, it is found that:

$$\begin{cases} E_a = 2M_+ \bar{\psi}_a e^{j(\omega t - \beta z)} \cos(Cz) \\ E_b = 2jM_+ \bar{\psi}_b e^{j(\omega t - \beta z)} \sin(Cz) \end{cases} \tag{11.2}$$

Note in passing that the phase change of $+ \pi/2$ which corresponds to a factor j in the expression for E_b. Guide B has a phase $\pi/2$ in advance of that of guide A. The energy flux associated with each elementary guide is:

$$P_a(z) = \cos^2(Cz) \tag{11.3}$$
$$P_b(z) = \sin^2(Cz) \tag{11.4}$$

Hence, the total power is equal to unity for all z.

The unit power associated with guide A for $z = 0$ is associated with guide B for $z = \pi/(2C)$. The length Λ is the repetition length or the beat length in which all the power is transferred from one medium to the other and then returned to the first:*

$$\Lambda = \frac{\pi}{C} = \frac{2\pi}{\beta_+ - \beta_-} \tag{11.5}$$

* Numerous authors call the length $L_c = \Lambda/2$ the coupling length. In practice it can be of the order of 1–2 cm.

124

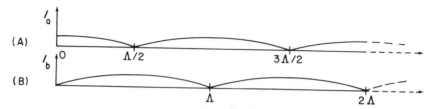

Fig. 11.2. Schematic representation of the passage of light between two identical parallel monomode guides coupled by evanescent waves. At $z = 0$, only guide A is fed with unit intensity. Coupling produces intensity $I_a \cos^2(C, z)$ in guide A and $I_b = \sin^2(C, z)$ in guide B. The length of the interval is $\Lambda = \pi/C$. Such a result is not evident, *a priori*. However, it is proved by calculation and experience

This is an approximate solution to the problem of coupling between two parallel, cylindrical monomode guides (Fig. 11.2).

Comparable results are obtained for the coupling of two parallel monomode guides of identical rectangular section and similarly for two monomode guides of which one has a circular section and the other a rectangular one. It is important that the polarizations are favourable and that the wave numbers β are very close. However, in the case where the sections are different, the power transmitted from one guide to the other can be only a small fraction of the total propagated power.

Coupling of parallel light guides recalls the coupling of two pendulums, of the same period, suspended from an elastic support. A free oscillation, imposed on one, transmits itself to the other and returns to the first.

If the photon in the rod can be represented by a particle of energy β^2, trapped in a potential well and subjected to a centrifugal force this particle can move to a neighbouring potential well by the tunnel effect and, conversely, return to the first, effectively oscillating continuously between the two.

11.3 MODULATORS AND DIRECTIONAL COUPLERS USING INTERFERENCE BETWEEN TWO GUIDES
(Practical switching devices for monomode optical guides)

In practice, to control coupling of monomode guides a phase variation $\Delta\varphi$ is induced in at least one of the guided waves by means of Pockels's effect. To this end, a transverse electric field is applied for which the electrodes are placed somewhere in the crystalline optical propagating medium at a distance h from each other. If L is the length of the guiding material in the direction of optical propagation:

$$\Delta\varphi = \pi \frac{V}{V_0} \frac{L}{h} \tag{11.6}$$

where L/h is of the order of 10^{-3} or 10^{-4}, V_0 is a characteristic voltage and V is of the order of 1 V.

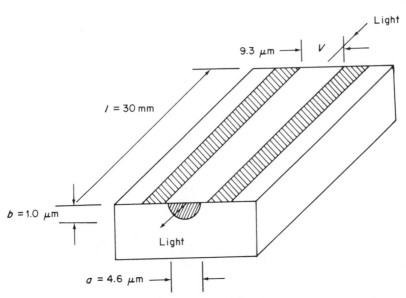

Fig. 11.3. Simple phase modulation. The ribbon guide, 4.6 μm wide and 1 μm thick, is formed by diffusing titanium into lithium niobate. On the surface, two longitudinal electrodes, spaced by 9.3 μm, modify the local value of refractive index when a potential difference V is applied. The result is a phase shift proportional to V. The band over which modulation is produced can reach several GHz. (After Kaminov *et al.*)

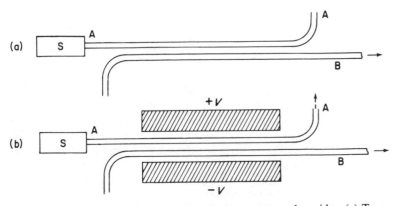

Fig. 11.4. Switching of the light between two monomode guides. (a) Two monomode guides A and B are parallel and close. Only guide A is fed with light. Coupling by evanescent waves exists between the two guides and all the light can pass from guide A to guide B, if the coupling length L is an integral multiple of a certain beat length L_c. (b) Using an electric field, to which both guides are subjected, the wave number β and hence the critical length L_c can be varied in such a way that light is switched either to guide A or to guide B. In another arrangement, electrodes allow β to be varied in one of the guides but not the other, which destroys the phase correspondence. Hence the light remains in guide A

126

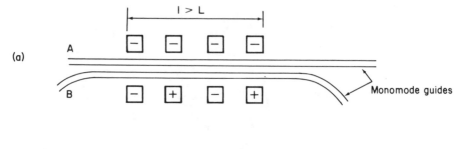

(a)

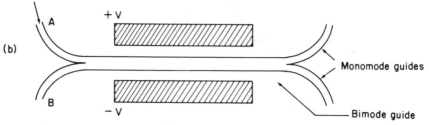

(b)

Fig. 11.5. (a) Two close, parallel ribbon monomode guides are coupled by evanescent waves if the phase propagates at the same velocity in the two guides. Application of a series of alternating voltages (4–12) allows the critical length L_c and the coupling length L of the system to be adjusted and allow light to pass, or not, from guide A to guide B. (b) A guide with two modes is substituted to couple the monomode guides. Variation of the voltage causes the light to arrive either at the upper or lower branch of the output of the two mode guide

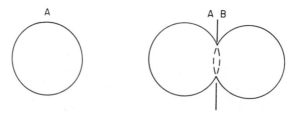

Fig. 11.6. The dielectric rod A surrounded by an exterior medium forms a monomode guide. The dielectric cylinder AB, whose cross-section consists of two intersecting circles of the same radius as A, surrounded by an external medium, constitutes a guide with two fundamental propagation modes. If the left part of AB is connected to monomode guide A, the light which was constrained by interface A will pass, almost entirely, from the left part of AB to the right part and then return to the left part over a length of $2L_c$. The light propagated in the external medium is, literally, attached to the light propagated in the rods

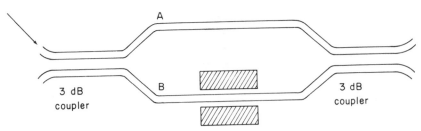

Fig. 11.7. Mach–Zehnder coupler. The coupler-switch is formed from two identical monomode guides. The first coupler, on the left, of length $\Lambda/4$ provides transfer of half the optical power which enters A to guide B. In guide B, the light has a phase lead of $\pi/2$. Further along, a second coupler of length $\Lambda/4$ can cause either transfer into B of the second half of the light if the phase there still leads by $\pi/2$ or return to A of the light in B if the phase there lags by $\pi/2$. Operation of the switch depends on an electro-optical phase modulator

An example, perhaps the simplest, of an electro-optic modulator is represented in Fig. 11.3. The measured energy required for operation is 6.8 μW/MHz of the band modulated. This has been described by Kaminov.[9]

In the COBRA configuration, described by Papuchon et al.[9] and approximately represented in Fig. 11.4(b), the guides are ribbons of $LiNbO_3$, and the arrangement is such that the field causes an increase in phase velocity in one of the guides and a reduction in the other. Finally, the most developed form of this type of modulator could be that represented in Fig. 11.5(b), which is due to Papuchon et al.

The Mach–Zehnder component, represented in Fig. 11.7 requires couplers which are produced for exactly 50% splitting in order to obtain good extinction. This modulator requires 1.0 mW/MHz.

As has already been indicated, coupling of light guides by evanescent waves can be used for modulation, multiplexing and switching. Integrated optical components based on coupling principles consequently have a vast field of applications.

The simple phase modulator presented in Fig. 11.3 is really a wide band modulator for optical guides. With this component, either an on/off amplitude modulator or a 2 × 2 switch (two inputs and two outputs) can be obtained.

11.4 DIRECTIONAL COUPLER MODULATORS USING DIFFRACTION GRATINGS

Examples will be presented here of a plane electro-optical modulator and an acousto-optic modulator in a ribbon guide. Notice, however, that there are phase and polarization modulators which use reverse-biased p–n double heterojunctions.

An electro-optic modulator proposed by Hammer and Phillips consists of a series of electrodes placed on the surface of a plane waveguide (see Fig. 11.8).

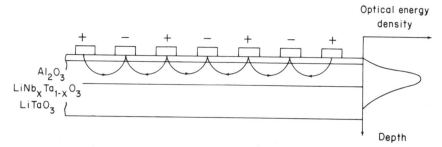

Fig. 11.8. A planar guide consists of a layer formed by diffusing niobium into lithium tantalate in order to form a crystal composed of $\mathrm{Li\,Nb}_x\mathrm{Ta}_{1-x}\mathrm{O}_3$ whose refractive index is greater than that of lithium tantalate and alumine. On the upper face a series of electrodes of alternate polarity (an interdigital comb) is aligned on the guide. When the voltage is applied, it produces a repetitive structure of refractive indices in the guide. A light wave incident at the Bragg angle is diffracted in the plane of the waveguide

The electric field produced by the electrodes, which are aligned serially on the guide and excited with an alternating voltage $(+V, -V)$, creates a repetitive pattern of refractive index values in the guide which are different from the normal value n. This structure performs the function of a diffraction grating. It causes deviation of the guided optical wave and could also cause reflection. If light is incident on the grating at less than the Bragg angle (θ) it is diffracted in the plane of guidance (at an angle 2θ). With a crystalline guide composed of lithium tantalate and niobate at $\lambda_0 = 6328$ Å, an extinction ratio of 20 dB is obtained with $V = 4V$ and 1.0 mW/MHz of bandwidth.

An acousto-optic light guide modulator is based on the interaction of a surface acoustic wave and a beam of guided light. An acoustic wave, produced by an interdigital comb, forms a refractive index pattern (of optical wave phase velocity) which is repetitive along a rectilinear path, in the homogeneous, generally crystalline and piezoelectric, material which guides it. A moving diffraction grating is therefore created in the acoustic waveguide which travels at the speed of the acoustic wave.

However, the optical guide superficially crosses the acoustic guide at the Bragg angle and at a small depth where the acoustic energy, which decreases exponentially with depth, is still substantial. The optical wave is diffracted, as by a stationary grating, if the velocity of the acoustic wave is small compared with that of light in the material $(v_a \ll v)$.

The modulator represented in Fig. 11.9 was proposed by Schmidt and Kaminov. The optical waveguide, formed as a Y, is in LiNbO_3, diffused with Ti to 500 Å. b is estimated at $2\,\mu\mathrm{m}$. For $f_a = 175\,\mathrm{MHz}$, $v_a = 3.3.10^5\,\mathrm{cm/s}$ and $\Lambda = 18\,\mu\mathrm{m}$. For modulation depth η of 70% and a bandwidth Δf of 30 MHz the electric power is 50 mW and the acoustic power $P_a \leqslant 8\,\mathrm{mW}$. Hence, $P_e/\Delta f = 1.7\,\mathrm{mW/MHz}$ and $P_a/\Delta f = 0.27\,\mathrm{mW/MHz}$.

(*Note*: Recently, Favre and Rivoallan have proposed an electro-optical amplitude modulator which acts directly on the refractive index profile of the guide

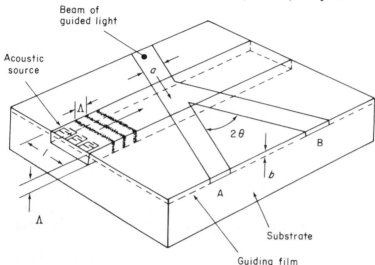

Fig. 11.9. A piezoelectric transducer produces a guided acoustic wave, of length Λ, which crosses a ribbon optical guide creating a diffraction grating of period $\Lambda/2$ in the material by variation of the refractive index. The light incident at the Bragg angle θ is deviated by 2θ in the direction of guide B.
(After J. M. Hammer[9])

and leads directly to amplitude modulation. The modulation depth obtained is 13 dB up to 300 MHz with $V = \pm 7.5$ V.)

11.5 CONCLUSIONS

Integrated optics continues with research and development into components suitable for modulation and switching of light. It has the classical advantages of miniaturization and integration and the power and control voltage are small. With monomode guidance, the available bandwidth has become very wide. The functions and processes which have been mastered are increasing remarkably.

Performance often depends on the quality of the crystalline or amorphous material used, and research into new materials and procedures to perfect production is being pursued.

12 Digital transmission

12.1 GENERAL

Optical fibre communication systems are principally applied to digital transmission using pulses. In a digital link, the signals can transmit any form of communication (the human voice, a written letter, an optical image, music, numerical data, etc.) in an appropriate standard form.

Transmission equipment is both less voluminous and less complex for a digital format than for an analogue one. For optical fibre links, pulse modulation is used, preferably in a digital format.

All forms of pulse modulation depend on Shannon's sampling theorem: 'A signal of limited bandwidth W can be reconstructed, without any distortion, from samples taken at a rate at least equal to twice the bandwidth of the signal.' (If the bandwidth is 1 MHz, the number of samples must be at least 2.10^6 per second.) The values of successive samples could be transmitted in an analogue form which would be pulse amplitude modulation. They could also be transmitted in digital form using a conversion scale, which is not necessarily linear, and a coding system with encription if required in order to ensure privacy of the transmission. This is termed 'pulse code modulation'. Coding of a sample can use the increment with respect to its predecessor. This is called 'delta modulation'.

It is possible, therefore, to change from an analogue format to a digital one during processing or transmission of a signal.

12.2 DIGITAL PULSE MODULATION (Fig. 12.1)

The values of samples can be coded in a linear scale with 2^n levels. If n is equal to 7 or 8 ($2^8 = 256$) the corresponding quantization error causes imperceptible effects on the reconstructed signal.

Transmission of a signal of bandwidth W by means of encoded numbers called 'codes' is clearly possible. This modulation is called 'pulse code modulation' and abbreviated to PCM (see Fig. 12.1).

The coded signal consists of a digital sequence of zeros and ones which represent the samples of the uncoded signal or message in a clearly defined manner.

130

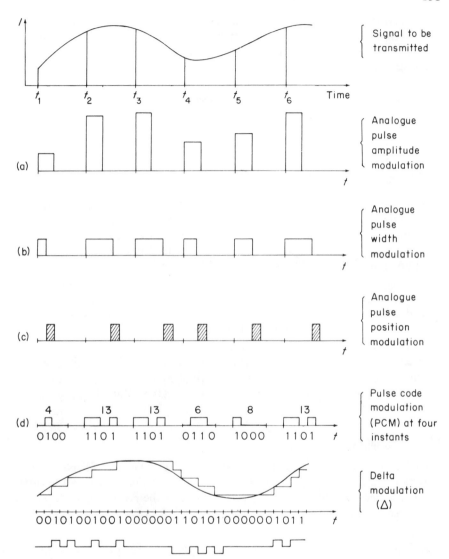

Fig. 12.1. Sampling and different types of pulse modulation for the same transmitted signal

When the code uses a differential measure of the amplitude of the sample (the difference between the amplitudes of present and previous samples) with a single level of increment (Δ), sampling must be very rapid. In recent systems, the value of Δ varies as a function of a parameter of the uncoded signal; for example, the slope; Δ^- modulation is successfully based on this principle.

The digital sequence obtained with PCM or delta modulation has the same structure, an apparently random sequence of ones and zeros. This digital sequence can be combined with other sequences. The combination of these sequences can be used to modulate the amplitude, frequency or phase of a carrier with the signal at high speed. This mode of operation is called 'multiplexing'.

The common characteristic of all digital modulation using impulses of light intensity transmitted by optical fibres is that every $1/b$ seconds (where b is the rate in bit/s) the receiver must choose between two, and two only, possible signals; a 'one' or a 'zero'.

12.2.1 Example of PCM

A telephone signal occupies the band 300–3400 Hz. For sampling, a frequency of 8000 Hz is chosen, a little greater than 6800 Hz. Sampling therefore occurs 8000 times per second. The analogue signal, represented as a continuous function, is replaced by a digital sequence consisting of a finite number of terms in unit time; one number every 125 μs.

In each interval of 125 μs the numbers corresponding to 30 independent telephone channels can be combined or multiplexed. This is multiplexing by time sharing or time-division multiplexing. The magnitude of each sample of the signal is expressed as an eight-digit binary number which produces eight electrical pulses of full or zero magnitude.

In each group of 30 telephone channels two supplementary channels are inserted, one for synchronization and one for signalling. Finally, the digit rate to be transmitted is 32×8 bits in 125 μs, that is, 2.048 Mbit/s (see Fig. 12.2).

12.2.2 Digital hierarchy

The CCITT (Comité Consultatif International Télégraphique et Téléphonique— International Telegraph and Telephone Consultative Committee) of the International Telecommunications Union has specified the norms for multiplexing digital sequences and has defined a hierarchy for Europe:

Order →	1	2	3	4	5
Rate (Mbit/s)	2.048	8.448	34.368	139.264	(560–840)
Capacity (channels)	30	120	480	1920	—

Telephone transmission on optical fibres is exclusively oriented towards normalized rates: 2, 34 and 140 Mbit/s. Data transmission (industrial and military) does not follow these norms: 100 kbit/s, 1 Mbit/s, etc.

At present, only transmission by fibre optics is assured; signal processing, sampling, multiplexing, analogue/digital conversion (with encription if required) and the inverse operations are achieved by electronic means.

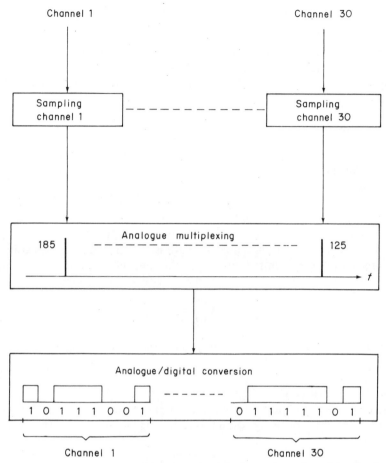

Fig. 12.2. Time-division multiplexing of 30 telephone channels and details of the composite digital signal

12.3 PRINCIPAL CHARACTERISTICS OF A DIGITAL LINK

12.3.1 The transmission code

For the receiver to read the received binary elements it is necessary to locate them conveniently within a time base; in other words, a clock must be provided which permits demultiplexing, reading and digital/analogue reconversion. The clock information could be transmitted separately. It is simpler to combine the binary information directly, and for this purpose, the digital signal is transmitted in a complex form which allows the clock timing to be recovered. A transmission code or 'line code' ensures correspondence without ambiguity.

12.3.2 Pulse regeneration

The receiver detects and amplifies the received pulses and re-forms them in synchronism with the clock. The clock timing is extracted from the coded information by means of a tuned circuit or phase-locked loop. The clock allows the appropriate times to be fixed and to decide whether the received pulse is a 'one' or a 'zero'.

12.3.3 The error rate

Spreading of the received pulse, due to pulling out of the time slot assigned for its reception and random parasitic noise, produces errors. In its most elementary form, the error arises from confusion between a 'light' and a 'dark' signal.

The quality of the reconstructed message is characterized by the error rate or the error probability, which is the ratio of the number of false bits and the total number of bits received, for a long sequence of pulses.

It can be shown that if the signal strength is a Gaussian random variable of standard deviation σ_u, centred on u when the transmitted symbol is '1' and on 0 when the transmitted symbol is '0', the probability of error is:

$$P_e = \frac{1}{2}\text{erfc}\left(\frac{Q}{\sqrt{2}}\right) \quad \text{with} \quad Q = \frac{u}{2\sigma_u} \tag{12.1}$$

on condition that the error probability is the same for each symbol

$$P_e = P(0/1) = P(1/0) \tag{12.2}$$

The threshold is equal to half the peak amplitude $u_s = (1/2)u$.

The parameter Q is proportional to the signal-to-noise ratio (S/N) which is defined as the ratio of the maximum received signal strength and the effective noise level:

$$Q = \frac{u}{2\sigma_u} = \frac{1}{2}\frac{S}{N} \tag{12.3}$$

In American literature, the notation $K = Q^2$ is often used:

$$P_e = \tfrac{1}{2}\text{erfc}(\sqrt{(K/2)}) \tag{12.4}$$

For $K \gg 1$, asymptotic expansion of the complementary error function allows it to be written as the first term:

$$P_e \simeq \frac{e^{-K/2}}{\sqrt{(2\pi K)}} \tag{12.5}$$

P_e	10^{-6}	10^{-7}	10^{-8}	10^{-9}	10^{-10}
Q	4.76	5.2	5.61	6	6.36
K	23	27	32	36	40

Values of Q and K for commonly encountered error rates.

12.4 TELECOMMUNICATION TRANSMISSION CODES

A digital transmission code for optical fibres must occupy as small a spectral width as possible to permit eventual multiplexing of several independent carrier wave-lengths and also to reduce the bandwidth of the receiver. This spectral width, or base band spectrum, should preferably become zero at zero frequency to avoid technical difficulties and to allow low-frequency components to be recovered which would disappear in the coupling capacitors of the receiver.

Furthermore, the code must allow the clock times to be recovered. The code must be adapted for binary messages and, finally, allow monitoring of the error rate.

12.4.1 Binary coding

The information to be transmitted is already in binary form. Representation of the two symbols 0 and 1 must be defined by two electrical signals (see Fig. 12.3).

12.4.1.1 NRZ coding (non-return to zero)

This is the first which comes to mind. A constant level corresponds to each symbol:

The level 0 for the symbol '0' $\Big\}$ during the interval T
The level A for the symbol '1'

where T is the basic interval or slot of the time base.

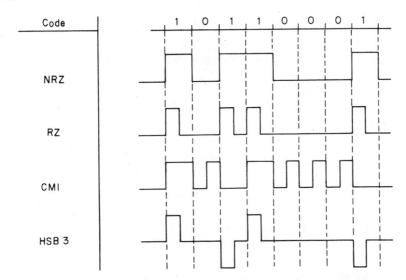

Fig. 12.3. Binary coding: NRZ (non-return to zero); RZ (return to zero) 'one' is represented by '10'; CMI (coded mark inversion) 'zero' is represented by '01' and 'one' alternately by '11' and '00'; HSB3 (high-speed bipolar 3) 'zero' is represented by '00' and 'one' by symmetrical $\pm$ '10' pulses, alternately positive and negative

The periodic signal $(0, 1)$ repeated indefinitely represents a well-known Fourier series expansion of which the principal term is at zero frequency.

If the binary signal is random, with the same probability of the two states, the power spectral density is of the form:

$$\Gamma(v) = A^2 T \left(\frac{\sin \pi T v}{\pi T v} \right)^2 \tag{12.6}$$

The function Γ has a zero for $v = 1/T$ and 90% of the energy is contained in the band $0.86/T$. It is necessary to recover the timing at the receiver.

12.4.1.2 RZ coding (return to zero)

This follows from the previous signal. The level A is maintained only in the interval $T/2$. Hence the name RZ. The periodic signal $(1,1)$ represents a Fourier series expansion which can be obtained from that for the NRZ periodic signal $(0, 1)$ by replacing T by $T/2$. It is clear that the spectral space occupied by RZ is twice that of NRZ, and 90% of the energy is contained in the band $1.72/T$.

In spite of the fact that these two codes have a maximum continuous component and do not transmit timing information, in view of their simplicity they are often employed in the USA and in Japan.

12.4.1.3 CMI coding (coded-mark-inversion)

This is without doubt one of the highest performance codes and one of the cleverest:

(1) The signal 01 is produced for each binary 'zero'. The element 01, which consists of the zero level during $T/2$ and level A during $T/2$, is generated by the inverse of the clock.
(2) The signal 11 or 00 is produced alternately for each binary 'one'. To cause the alternation, a memory state must be changed. For decoding, the transition times of the signal are detected. If a transition occurs at one half the bit time $(kT + T/2)$, a 'zero' has been received. Otherwise, a 'one' has been received.

Negative transitions are in phase with the clock and allow the timing to be recovered. A resonant circuit at the clock frequency can be used for this purpose; a low magnification factor can be used since the interval between two negative transitions is $2T$ at maximum.

This code does not transmit energy at zero frequency. It has one clock band in the spectrum and it allows error detection: the combination 10 should not occur and sequences 00 and 11 must alternate. The bandwidth at 90% energy is $1.70/T$.

12.4.1.4 HSB3 coding (high-speed, bipolar, 3)

This is a three-level code (it allows connection to a ternary channel):

(1) A continuous signal at the zero level is produced for each binary 'zero'. However, each block of four zeros is systematically replaced by a particular sequence. Hence there are never more than three consecutive zeros.
(2) Symmetrical pulses of duration $T/2$ alternating between positive and negative are produced for each binary 'one'. No energy is transmitted at zero frequency and the bandwidth at 90% energy is about $1.5/T$.

If the binary signal is random, with the same probability of the two states, the power spectral density is of the form:

$$\Gamma(v) = \frac{A^2 T}{4} \left(\frac{\sin \pi T v/2}{\pi T v/2} \right)^2 \sin^2 (\pi T v) \tag{12.7}$$

Fibres are not made for multilevel codes but for the alternatives of light and dark. Light emitters introduce non-linearities which can make the level differences unequal. There are numerous other codes.

12.5 DIGITAL RECEPTION

On transmission, the electrical signal has been transformed into an optical one whose amplitude represents a power. On reception the optical signal is again transformed into an electrical one whose power is proportional to the square of the amplitude.

The conventional performance measure of an optical receiver is the minimum detectable power (MDP). This quantity depends, among other things, on the noise in the receiver and the bit rate. For a given bit rate and a given error rate, the minimum detectable power also depends on the transmission code.

12.6 EXAMPLE OF ANALYSIS OF DIGITAL RECEPTION

Assume that the code employed is NRZ, which is currently used in small systems. Let the pass band B of the receiver be provisionally equal to the bit rate b:

$$B(\text{Hz}) = b(\text{bit/s}) \tag{12.8}$$

Also let:

$$b = \frac{1}{2T} \tag{12.9}$$

where T is the duration of the transmitted pulse (see Fig. 12.4). The binary signal consists of two signal elements, which are the presence or absence of light and which are represented by '1' or 'mark' and '0' or 'space'. The signal appears as a current of peak value $\hat{i}$.

The decision level will not be optimized and the decision threshold will be

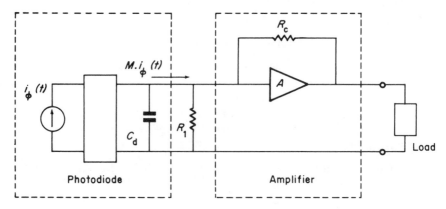

Fig. 12.4. Transconductance receiver, equivalent circuit

taken as equal to:

$$(1/2)\hat{\imath} = I_{\text{thr}} \qquad (12.10)$$

If $i_\phi(t) \geqslant I_{\text{thr}}$, the signal is '1'; in the other case it is '0'. Now assume that the detected signal is a Gaussian random variable of standard deviation σ_i, centred on $2I_{\text{thr}}$ for the symbol '1' and on 0 for the symbol '0'.

Assume an error probability of 10^{-9}, that is, $K \simeq 36$. Consequently $Q = 6$ and the signal-to-noise ratio $S/N = 12$ or 10.8 dB. The two major sources of noise are:

(1) Quantum noise proportional to the detected current; and
(2) Amplification noise, which includes thermal agitation noise proportional to temperature and noise produced by the active components, such as transistors, which can be represented by a current source and a noise voltage source.

The quantum noise is:*

$$\langle i_q^2 \rangle = 2q(I_\phi)M^2F(M)\cdot\Delta f \quad \text{where} \quad I_\phi \equiv \langle i_\phi(t) \rangle \qquad (12.13)$$

Now:

$$I_\phi = \frac{\eta q}{h\nu}P_0 \quad \text{where} \quad P_0 \equiv \langle p(t) \rangle \qquad (12.14)$$

From which:

$$I_\phi = S_e \cdot P_0 \qquad (12.15)$$

and:

$$\langle i_q^2 \rangle = 2qS_e \cdot M^2 \cdot F(M)P_0 \cdot \Delta f \qquad (12.16)$$

Here Δf will be taken as equal to the bandwidth B.

* See the notation of Chapter 10, page 115.

12.6.1 Thermal noise

A transconductance amplifier with feedback resistance R_c is chosen (see Fig. 12.4). The thermal noise current can be written:

$$\langle i_{th}^2 \rangle = \frac{4kT \cdot B}{R_c} \cdot F_{th} \qquad (12.17)$$

Where T is the absolute temperature and F_{th} the thermal noise factor of the receiver.

In the presence of thermal noise alone, the minimum detectable optical power (MDP) is obtained when:

$$I_\phi = \frac{\sqrt{\langle i_{th}^2 \rangle}}{M} \qquad (12.18)$$

Hence:

$$\sqrt{\langle i_{th}^2 \rangle} = M \cdot S_e \cdot (\text{MDP}) \qquad (12.19)$$

The equivalent optical power is proportional to $\sqrt{B}$.

To allow numerical comparison of the quality of receivers, they are assessed by means of the equivalent thermal noise power (ENP).*

$$(\text{ENP}) = (\text{MDP})/\sqrt{B} \quad (\text{watt}/\sqrt{\text{Hz}}) \qquad (12.20)$$

or:

$$(\text{ENP}) = \sqrt{\left(\frac{4kT}{R_c} F_{th} \right)} \Bigg/ S_e \cdot M \qquad (12.21)$$

The total noise current is:

$$\langle i_q^2 \rangle + \langle i_{th}^2 \rangle = 2q \cdot S_e \cdot M^2 \cdot F(M) \cdot P_0 \cdot B + S_e^2 M^2 [(\text{ENP})\sqrt{B}]^2 \qquad (12.22)$$

At the amplifier input, the signal current is:

$$i_s(t) = M i_\phi(t) \qquad (12.23)$$

and:

$$\overline{(i_S(t) - I_{thr})^2} = M^2 \cdot S_e^2 \overline{(p(t) - P_0)^2} \qquad (12.24)$$

Finally it is found that

$$K = \frac{\overline{(i_S(t) - I_{thr})^2}}{\langle i_q^2 \rangle + \langle i_{th}^2 \rangle} = \frac{P_0^2}{2\dfrac{h\nu}{\eta} P_0 \cdot F(M) \cdot B + [(\text{ENP})\sqrt{B}]^2} \qquad (12.25)$$

and

$$P_0 = KF(M)\frac{h\nu B}{\eta} \left\{ 1 + \sqrt{\left[1 + \frac{1}{K} \left(\frac{(\text{ENP})}{F(M)(h\nu/\eta)\sqrt{B}} \right)^2 \right]} \right\} \qquad (12.26)$$

* The thermal noise power equivalent to the power of the minimum signal acceptable.

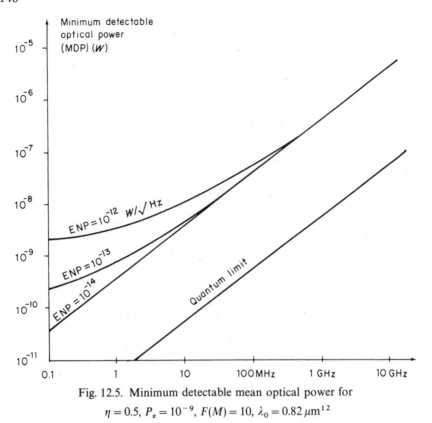

Fig. 12.5. Minimum detectable mean optical power for

$$\eta = 0.5, \; P_e = 10^{-9}, \; F(M) = 10, \; \lambda_0 = 0.82 \, \mu m^{12}$$

This expression assumes that the thermal noise of the receiver is effectively provided by the ENP.

Figure 12.5 shows the minimum detectable optical power for $K = 36$, that is, $P_e = 10^{-9}, \eta = 0.5, F(M) = 10$ and $\lambda_0 = 0.82 \, \mu m$. The slope is proportional to the bandwidth.

The quantum limit is less than:

$$P_0 = 2K \cdot F(M) \frac{h\nu B}{\eta} \tag{12.27}$$

since Gaussian fluctuations are assumed.

The quantum limit calculated for $P_e = 10^{-9}$ is given by:

$$P_{\min} = 10 \frac{h\nu B}{\eta} \tag{12.28}$$

These results can be improved by optimizing the receiver, first by choosing a bandwidth B suited to the bit rate b. Briefly, it has been established that a transconductance amplifier following an avalanche photodiode serves mainly to reduce the rise time of the pulse.

13 Particular applications

Optical fibres have numerous applications other than the transmission of telecommunication signals which depend on particular or novel characteristics of optical fibres that are not possessed by metallic guides.

13.1 THE OPTICAL FIBRE IS AN INSULATOR

The optical fibre protects against electrocution, and transmission of signals is practically insensitive to surrounding em fields, even very intense ones, which are not disturbed by the presence of the optical fibre. Consequently, an optical fibre link can be used:

(1) In the field near to an antenna without disturbing it;
(2) To measure very high voltages on power transmission lines;
(3) In avionic links, to protect against storm discharges in flight; and
(4) In armament systems, to protect against nuclear electromagnetic pulses (EMP).

13.2 THE INTENSITY OF THE LIGHT EMERGING FROM A GIVEN DIELECTRIC ROD IS SENSITIVE TO THE REFRACTIVE INDEX OF THE EXTERNAL MEDIUM

Guidance of light in a glass rod of refractive index n_1, rectilinear or 'U' shaped, is very sensitive to the value of the refractive index n_2 of the external medium; in this context the parameter Δ, called the 'relative deviation' of the indices, has been defined. The quantity of light propagated is approximately proportional to:

$$\Omega_{ACC} = 2\pi\{1 - \sqrt{[1 - (n_1^2 - n_2^2)]}\} \quad n_1 > n_2$$

At the extremity of the rod the optical energy reflection coefficient of the interface also depends on Δ. Consequently, an interconnected glass rod can be used to observe or measure a change in the exterior medium, which is represented by a change in its refractive index. This method is recommended whenever use of an electric current could be dangerous (Fig. 13.1).

141

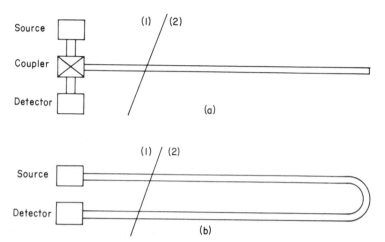

Fig. 13.1. Principle of operation of equipment to detect and measure a variation in the refractive index of a liquid or gaseous medium exterior to a dielectric rod

Since the refractive index of glass is about 1.5, water about 1.33 and air 1, small systems can be envisaged which, for example, permit detection of the presence of water instead of air in a mineshaft or of vapour instead of liquid in a reservoir. By using a source of light, which operates in an absorption band of methane, the presence of firedamp could also be detected. Finally, it is possible to observe and measure the variation of absorption of light by liquid such as sea water or the electrolyte of a battery. In this case, the presence of numerous ions in the water introduces an extinction index which reduces the reflected and guided light.

13.3 THE PHASE OF THE LIGHT EMERGING FROM AN OPTICAL GUIDE ALLOWS VARIATION OF THE PHYSICAL CHARACTERISTICS OF THE ENVIRONMENT TO BE OBSERVED

In discussing propagation conditions of light in a guide it has been assumed that the materials of the media which form the guidance structure are permanent, that is, independent of time and the value of physical characteristics of the environment (temperature, pressure, electrical and magnetic fields), which can vary with time. This assumption is quite permissible since glass and vitreous silica in particular have a remarkably constant behaviour in a perturbed environment.

However, light can be observed in many ways, and such precision can be achieved that a small change can reveal variation of a characteristic of the external medium and allow it to be measured, if the length of the fibre is sufficient and comparison with a stable reference is possible.

Consider, for example, a light source of wavelength $\lambda_0 = 820 \, nm$ and a monomode fibre 20 m long. The wavelength in the guide is:

$$\lambda_z = \frac{2\pi}{\beta} \simeq \frac{2}{3} \cdot 820.10^{-9}\,\text{m}$$

The number of wavelengths within the guide is $0.37.10^8$. If a phase shift of 1 radian can be observed, a variation in refractive index equivalent to $c.\beta/\omega$ or $0.43.10^{-8}$ can be detected.

The sensitivity evidently arises from the length of the guide. Thus it is possible to measure magnetic field, an electric field or a velocity of angular rotation.

13.3.1 The fibre optic hydrophone

An acoustic sensor consisting of a monomode fibre is immersed in the water and an identical parallel fibre is coupled to the first by two 3 dB beam separators but not subjected to the acoustic wave. The two light waves at the output have approximately the same amplitude but a different phase. A change in pressure exerted on the sensor is represented by a shift in the relative phase. Interference between the two light waves, one a fixed reference and the other phase modulated by the acoustic pressure, is produced on the surface of a photodetector. In Figure 13.2 it can be seen that modulation of the reference frequency is provided for heterodyne reception if required.

An important difficulty arises from the fact that a pressure wave which acts on the rod modifies not only the diameter but also the length, and the two effects oppose each other. To remedy this situation, the fibre can be surrounded by a plastic coating which has a tendency to elongate under the effect of increased

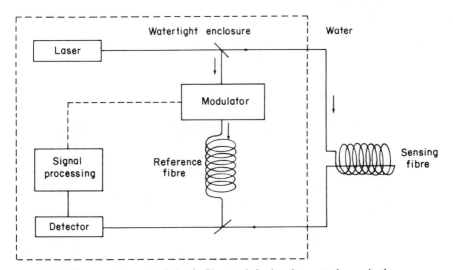

Fig. 13.2. Diagram of the principle of a fibre optic hydrophone. A change in the pressure exerted on the sensor leads to a variation in the relative phase of the two light beams which arrive at the surface of the photodetector

144

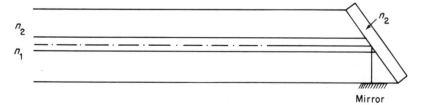

Fig. 13.3. Critical angle acoustic detector. A fibre with core refractive index n_1 and cladding index n_2 is terminated by a plate of index n_2. It operates in monomode. The incidence angle on the plate at the extremity is just above the critical value. Reflection is therefore total and a small mirror allows return of the light to the input. If an excessive pressure, caused, for example, by an explosion, changes the values of n_1 and n_2 differently, total internal reflection will no longer occur. The light returning to the input will suddenly drop. Control is difficult unless a source of variable wavelength is used

pressure; this will reduce the shortening of the fibre without significantly modifying the variation in refractive index:

$$\Delta\phi = k_0 \left(\frac{dn}{dP} + \frac{n}{L} \frac{dL}{dP} \right) P \cdot L$$

where n is the equivalent refractive index, P the pressure and L the length.

In heterodyne reception the output current produced by a sinusoidal acoustic wave of angular frequency ω_s is:

$$i_s = G \Delta\phi_{max} \sin \omega_s t$$

where G is a constant and $\Delta\phi_{max}$ the maximum phase shift. With this type of equipment, an acoustic pressure of the order of a micropascal can be detected (Fig. 13.3).[7]

13.3.2 The fibre optic gyroscope

It is known that, within a closed space, uniform rectilinear motion with respect to a (Copernican) reference frame cannot be detected by any mechanical or electromagnetic means. (Note the word 'rectilinear' in this sentence.) On the other hand, it is also known that non-rectilinear and/or non-uniform motion of the surroundings or a vessel can be immediately detected by passengers without resorting to electromagnetism. Accelerometers have enabled inertial navigation systems to be developed.

However, an electromagnetic effect, described in 1913 by the French physicist Georges Sagnac, allows angular velocities of rotation to be measured and competes with mechanical gyroscopes. Sagnac's interferometer is a closed planar optical circuit (see Fig. 13.4(a)) in which monochromatic light propagates in air with velocity c, circulating in clockwise and anticlockwise directions. If the circuit is given a rotational movement about an axis perpendicular to its plane, the light which circulates in the sense of rotation of the system covers a greater distance than that which travels in the opposite direction. Consequently, interference between the two light beams will be modified, the phase change $\Delta\phi$ of one beam with respect to the other being proportional to the angular velocity Ω.

More precisely, if u is the tangential velocity of a circular path, the phase shift $\Delta\phi$ is, to a first order, proportional to u/c (see Fig. 13.4(b)).

If two beams of coherent light are propagated in opposite directions along the same fixed circular path, they will be in phase again when they return to their origin.

In contrast, if the guide and the light source rotate together with tangential velocity u, the beam which rotates in the same direction as the guide and travels at velocity c covers a much greater distance than the beam which rotates in the reverse direction, the difference being $(2\pi R/c).2u$.

After one rotation the phase difference between the two beams will be:

$$\Delta\phi = \frac{8\pi^2 \cdot R \cdot u}{c \cdot \lambda_0} \text{ (rd)}$$

These effects can be analysed differently by noticing that the frequency of a photon depends on the reference frame of the observer:

Consider a point source 0 on the circular path and a Galilean reference $0xyz$ which are fixed with respect to the guide. The $0z$ axis is tangential to 0 on the circular path and, to a first approximation, is in rectilinear uniform motion of velocity u with respect to the Copernican reference $0'x'y'z'$, which is derived from $0xyz$ by a translation $00'$ along $0z$. Light emitted at 0 in the moving reference frame has the same frequency in the increasing z direction as in the decreasing z direction.

In contrast, in the fixed reference frame the frequencies are given by the relativistic formula of the Doppler–Fizeau effect for the photon, and are:

$$v'_+ = v \sqrt{\left(\frac{1 + u/c}{1 - u/c}\right)} \quad \text{and} \quad v'_- = v \sqrt{\left(\frac{1 - u/c}{1 + u/c}\right)}$$

The phase difference between the two beams, returning after one rotation in the $0'x'y'z'$ reference is

$$\Delta\phi = \frac{4\pi^2 R}{c}(v'_+ - v'_-) = \frac{8\pi^2 R \cdot u}{c \cdot \lambda_0}$$

For a single plane circuit, of any form, enclosing area A, perpendicular to the angular velocity vector Ω, the calculated phase shift, neglecting infinitely small second- and higher-order terms, will again be:

$$\Delta\phi = \frac{8\pi \vec{A} \cdot \vec{\Omega}}{c \cdot \lambda_0}$$

For simple navigation it is necessary to achieve one-thousandth of the speed of

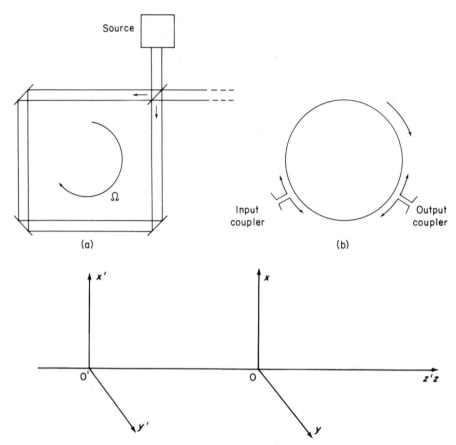

Fig. 13.4. (a) The principle of the interferometer proposed by Sagnac in 1913; (b) a circular fibre optic arrangement with N turns (at $\lambda_0 = 0.85\,\mu$m, a fibre of length $L = 10$ km includes 1.7×10^{10} wavelengths)

rotation of the Earth, which is 15 degrees per hour;

$$0.73.10^{-7} \text{ radian per second (rd/s)}$$

This has been achieved and fibre optic gyroscopes are starting to be used in aircraft.

One of the limitations is quantum noise of the photon. This fundamental limitation to sensitivity corresponds to an equivalent noise of the order of $10^{-8}\,$rd/$\sqrt{\text{Hz}}$ for the Sagnac interferometer and an optical power of several milliwatts.

Reciprocity assumes that the optical path is unique, that is, it operates in a unique mode. The two beams, in opposite directions, must conform exactly to the fundamental solution, precisely at the same instant, in order to obtain a reading within 10^{-6} of an interference fringe. At present, efforts are being made in two directions:

(1) Development of integrated optical technology with a view to mass production of reciprocal interferometric guides; and

(2) Research into the best polarization and modulation to limit quantum noise.

In the present situation distinction can be made between:

(1) The 'laser circuit gyroscope' or gyro-laser, in which the tube of a gas laser (HeNe, for example) forms an integral part of a triangular or rectangular circuit with three or four mirrors; and

(2) The 'fibre optic gyroscope', in which the optical circuit consists of a monomode fibre, from several hundreds to several thousands of metres long, wound into a coil.[6] With a silica fibre of 2–5 dB/km and 10 mW one could hope to detect 10^{-9} rad/s.

14 *Small systems*

14.1 INTRODUCTION TO THE FIRST TWO GENERATIONS
OF COMPONENTS

A 'small system' will refer to one without repeaters/regenerators and without switching, as opposed to large systems which use repeaters to cover long distances and switches to establish connections and modify traffic routes. Elementary systems will be considered which use optical fibres and a detector, which is an optical/electronic interface, and, if necessary, the optical source which is an electronic/optical interface.

At present there are two generations of these components. The first originated in the transparency window of silica which extends from the visible to the OH resonance at $0.95\,\mu m$. In this window, where the inevitable Rayleigh scattering prevails, the low kilometric attenuation which is encountered at longer wavelengths cannot be expected. This is called the 850 nm band.

LEDs and SCLs which use a GaAs/GaAlAs junction and silicon PIN and avalanche diodes are encountered; they originally allowed these wavelengths to be used.

When the distance to be covered is small, operation in this band and use of first-generation components must be considered. The limit is about 10 km. When it is required to cover distances of several kilometres, higher pass bands can be obtained.

The second generation of components takes advantage of fibres of low or very low OH content. These allow the use of longer wavelength, at which Rayleigh scattering becomes negligible, towards $1.2\,\mu m$.

This second band is called the 1200 nm band. Components for the first band cannot be used. (The GaAs/GaAlAs junction does not function above $0.9\,\mu m$ and silicon detectors do not function above $1.1\,\mu m$.)

The most commonly used components operate in the $1.0–1.6\,\mu m$ band. They use quaternary compounds (Ga, In)(As, P) which have a higher degree of freedom than ternary compounds. Ge, GaSb and InSb photodiodes are also used.

It is also possible to dope silica in such a manner as to cancel material dispersion at a frequency which can be freely chosen between 1 and $1.6\,\mu m$. For these reasons, the 1200 nm band is preferred for long distances and high bandwidths.

148

14.2 STUDY OF A SMALL PROJECT SYSTEM

Such a study involves the best choice of components and a specification of the characteristics of the system. It consists of collecting information on components which are available on the market and replying, in a successive order, to the following questions:

(1) *What is the optical power necessary for detection?* All depends on what is to be done with the detected signal. If it is a digital signal, the acceptable error rate must first be specified.
(2) *What is the optical power available at the input to the guide?* This depends on possible and available sources. It is necessary to take account of coupling losses.
(3) *How are the available decibels distributed?* From the replies to questions (1) and (2) a power attenuation expressed in decibels can be deduced. A choice must be made between several mono- or multifibre cables and the available decibels distributed between attenuation due to distance, interconnections, couplings, possible losses of optical energy and, finally, a safety margin. It is equally necessary to verify, in each case, that the bandwidth of the guide and its terminal components allow transmission of the signal.
(4) *What is the best conceivable system?* As the amount of information is large, research should be made with the assistance of a computer. It is necessary to obtain the best specification and to show that any modification involves an increase in overall cost.

14.3 DIGITAL APPLICATIONS

14.3.1 Example 1

For a link at $\lambda_0 = 0.90\,\mu$m, with digital pulse modulation, a signal-to-noise ratio $K = 32$ is specified, to obtain an error rate $P_e = 10^{-8}$. It is required to calculate the optical power necessary for detection, in the presence of quantum noise, for an avalanche photodiode having an excess factor $F(M)$ equal to 6. The bandwidth is 60 MHz.

Solution: The quantum noise is given by equation (12.16) in Chapter 12. Equations (12.25) and (12.26) give the result. By assuming that thermal noise is negligible compared with quantum noise, it becomes:

$$P_0 = K \cdot F(M) \cdot \frac{h\nu}{\eta} B \cdot 2 \quad \text{with} \quad h\nu = h\frac{c}{\lambda_0}$$

$$h \simeq 6.6\,10^{-34}\,(\text{J.s}) \quad \text{and} \quad c \simeq 3.10^8\,\text{m/s}$$

From which:

$$10\log P_0 = 16 - 260 + 10\log\frac{K \cdot F(M)}{\eta} \cdot \frac{B}{\lambda_0}$$

when quantities are measured in SI units.

Alternatively:

$$10 \log P_0 (\text{mW}) = -94 + 10 \log \frac{KF(M)}{\eta} \cdot \frac{B(\text{MHz})}{\lambda_0 (\mu m)}$$

It is found that:

$$10 \log P_0 = -52 \, \text{d Bm}$$

The number which gives the power in milliwatts is expressed in decibels.

Comment: For a PIN photodiode, $F(M) = 1$ would have been inserted in equation (12.26), to give

$$P_0 = K \frac{h\nu}{\eta} B.2$$

14.3.2 Example 2

For the same conditions as above it is required to calculate the power necessary for detection in the presence of thermal noise of the avalanche photodiode detector/amplifier with a $(\text{ENP}) = 5.10^{-14} \, (W/\sqrt{Hz})$.

Solution: Assume for this application that the quantum noise $\langle i_q^2 \rangle$ in equation (12.25) is negligible compared with $\langle i_{th}^2 \rangle$. It is found that:

$$P_0 = (\text{ENP})\sqrt{(K \cdot B)}$$

From which:

$$P_0(\text{mW}) = 10^6 \cdot (\text{ENP})(W/\sqrt{Hz}) \cdot \sqrt{[K \cdot B(\text{MHz})]}$$

and:

$$10 \log P_0(\text{mW}) = 60 + 10 \log [(\text{ENP})(W/\sqrt{Hz})] + 5 \log K \cdot B(\text{MHz})$$

With the above data this becomes:

$$10 \log P_0(\text{mW}) = -57 \, \text{dB}$$

Consequently, an avalanche diode can be chosen for the receiver which provides an equivalent thermal noise power of $-57 \, \text{dBm}$ or $5 \, \text{dB}$ less than that of quantum noise.

14.3.3 Example 3

Retaining the same conditions, it is required to establish a power assessment for an 8 km link at 45 Mbit/s with a bandwidth of 50 MHz and an error rate $P_e = 10^{-8}$.

Solution: The nominal power necessary at the receiver is $-52 \, \text{dBm}$. The attenuations are:

Coupling (transmitter)	$1 \times 3\,\text{dB}$	3
Connections	$2 \times 1\,\text{dB}$	2
Splices	$7 \times 0.3\,\text{dB}$	2.1
Cable	$8\,\text{km} \times 5\,\text{dB/km}$	40
Coupling (receiver)	$1 \times 1\,\text{dB}$	1
	Total attenuation:	48.1 dB
	Safety margin:	8.9 dB

The power necessary for transmission is:

$$-52 + 57 = 5\,\text{dBm}$$

A 5 mW laser diode would be selected.

(*Note*: A complete study should include a calculation of the bandwidth or, at least, a calculation of the rise time of the detector current, taking account of the response of the source and the optical guide.)

14.3.4 Example 4

It is required to establish a power assessment for a 32 km link at $\lambda_0 = 1.2\,\mu\text{m}$, which has a capacity of 34 Mbit/s and uses a graded index optical fibre; the power necessary at the receiver is $-42\,\text{dBm}$.

Solution: The attenuations are as follows:

Coupling (transmitter)	$1 \times 3\,\text{dB}$	3
Connections	$2 \times 0.85\,\text{dB}$	1.7
Splices	$5 \times 0.2\,\text{dB}$	1.0
Cable	$32\,\text{km} \times 1.1\,\text{dB/km}$	35.2
Coupling (receiver)	$1 \times 1\,\text{dB}$	1
	Total attenuation:	41.7 dB
	Safety margin	4.3 dB

The power necessary for transmission is:

$$-42 + 46 = 4\,\text{dBm}$$

A 4 mW laser diode would be selected.

14.3.5 Example 5

Calculate the bit rate of a graded index fibre which has a parabolic law ($g = 2$), a length of 32 km, an effective numerical aperture of 0.17 and a refractive index n_1 along the axis of 1.5. The value $N_1 = 1.6$ for the group index corresponds to the value n_1 for the refractive index.

Solution: It is known that (equation (5.7) $NA_{\text{eff}} = NA(0)/\sqrt{2}$. From which: $NA(0) = 0.17 \times 1.414 = 0.24 = n_1\sqrt{2\Delta}$ (equation (5.5)). Hence $2\Delta = 0.0257$.

The spread in trajectory duration caused by modal dispersion (see Chapter 5, equation (5.24)) is:

$$\sigma = 0.150 \cdot 4 \cdot N_1 \frac{\Delta^2}{c}$$

(for $g = g_0 = 2$ and a coefficient 0.150 instead of 0.144).

It is found that

$$\Delta^2 = 1.65 \cdot 10^{-4} \qquad \sigma = 4.25 \, \text{ns}$$

and:

$$b = \frac{1}{4\sigma} = 58 \, \text{Mbit/s}$$

14.3.6 Example 6

It is required to know the potential bit rate of a parabolically graded fibre of length 1 km, assuming $n_1 = 1.5, \Delta = 0.01$ and negligible material dispersion.

Solution: As above

$$\sigma = 0.150 \cdot L \cdot n_1 \frac{\Delta^2}{c}$$

(see Chapter 5, equation (5.24)). It is found that:

$$b = \frac{1}{4\sigma} = 3.3 \, \text{Gbit/s}$$

The potential capacity of such a fibre is:

$$b_0 = 3.3 (\text{Gbit/s}) \times \text{km}$$

14.3.7 Example 7

It is required to know the bit rate of a step index fibre of 1 km length, assuming $n_1 = 1.5$ and $\Delta = 0.01$.

Solution: The spread in trajectory duration caused by modal dispersion is:

$$\frac{\Delta t}{L} = \frac{n_1}{c} \sqrt{(2\Delta)}$$

(see Chapter 3, equation (3.19)). Hence for $L = 1 \, \text{km}$, $\Delta t = 50 \, \text{ns}$.

The bit rate is restricted to 10 Mbit/s if the data period is $2\Delta t$ and material dispersion is not taken into account. Over longer distances, taking account of practical results, leakage modes and trains of pulses (see Chapter 5, equation (5.23)) one has:

$$\sigma = 0.39 \frac{n_1 \Delta}{c} \cdot L$$

which gives, for 5 km, for example:

$$b = 2.6\,\text{Mbit/s}$$

14.3.8 Example 8

Consider a light-emitting diode whose energy intensity along the normal to the radiating surface is $I(0) = 3\,\text{mW/sr}$. What is the emitted power of this diode assuming that it is a Lambertian radiator?

Solution:

$$P_e = \pi \cdot I(0)$$

(see Chapter 9, equation (9.10)) from which:

$$P_e = 9.42\,\text{mW}$$

14.3.9 Example 9

Consider a fibre having a core diameter $2a = 100\,\mu\text{m}$ and a numerical aperture $NA = 0.25$. What is the power coupled to the fibre by a Burrus diode of the same diameter and luminance $L_e = 66\,\text{W/cm}^2/\text{sr}$?

Solution: The coupled power (see Chapter 9, equation (9.24)) is:

$$P_c = L_e \cdot A_c \cdot \Omega_{\text{ACC}}$$
$$P_c = L_e \pi a^2 \cdot 2\pi(1 - \cos\theta_{\text{ACC}})$$
$$P_c = 1.0\,\text{mW}$$

(see equation (9.1)). (The angle of acceptance is about 15 degrees and therefore sufficiently small to assume that the energy intensity is constant in the solid angle of acceptance.)

14.4 SUMMARY RELATING TO DISPERSION

14.4.1 Theoretical results

(1) For a step index fibre:

SI $\quad \sigma = 0.39\dfrac{N_1\Delta}{c} \cdot L$

(without coupling of modes)

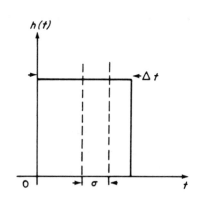

(2) For a graded index fibre:

GI $\sigma = 0.15\dfrac{N_1\Delta^2}{c}\cdot L$

(without coupling of modes)

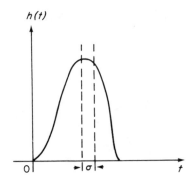

14.4.2 Measured results

(1) Frequency domain:
3 dB bandwidth B:

$B_0(\text{MHz}\cdot\text{km}) = \dfrac{190}{\sigma(\text{ns/km})}$

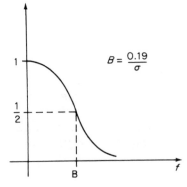

$B = \dfrac{0.19}{\sigma}$

(2) Time domain:
Time spread T at 3 dB:

$B_0(\text{MHz}\cdot\text{km}) = \dfrac{450}{T(\text{ns/km})}$
half-max

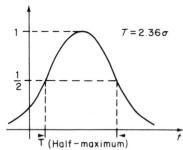

$T = 2.36\sigma$

T (Half-maximum)

14.4.3 Mean values not very variable in practice

Step index	Graded index
$B_0(\text{MHz}\cdot\text{km}) = 0.12\dfrac{1}{\Delta}$	$B_0(\text{MHz}\cdot\text{km}) = 0.40\dfrac{1}{\Delta^2}$
$B_0(\text{MHz}\cdot\text{km}) = 0.57\dfrac{1}{(NA)^2}$	$B_0(\text{MHz}\cdot\text{km}) = 8.13\dfrac{1}{(NA)^4}$
$T(\text{ns/km}) = 3700\cdot\Delta$ half-max	$T(\text{ns/km}) = 1125\cdot\Delta^2$ half-max

(*Note* 1: $NA = n_1 \sqrt{(2\Delta)}$.

Note 2: Manufactureres specify the product $B_0 = B.L$ or $T_{\text{half-max}}$ and the numerical aperture NA.)

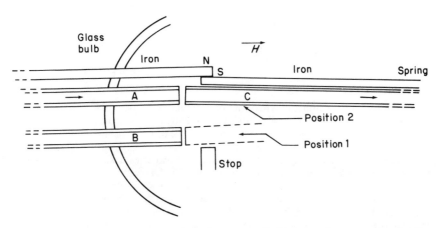

Fig. 14.1. Diagram of a fibre optic switch controlled by an electromagnet. In the absence of the magnetic field, fibre C is in position 1 against the stop coupled to fibre B. For fibres with $\phi = 125\,\mu$m the glass bulb has a diameter of 5 mm and a length of 25 mm. The switching time is from 1 to 2 ms. A pair of switches of this kind can allow remote control of the replacement of a faulty repeater by a spare

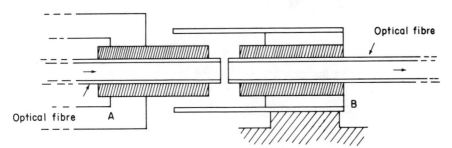

Fig. 14.2. Revolving optical joint. There is no direct mechanical contact. A is the rotating part at the extremity of an axis of rotation, B is the fixed part. Data in digital format will be transmitted from A to B more positively than by rotating electrical contact. Slight longitudinal and transverse movements are possible without affecting the transmission of optical signals

156

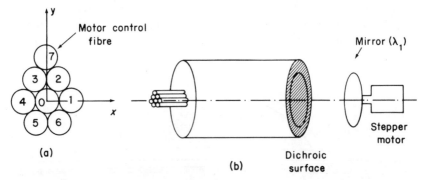

(a)

(b)

Dichroic surface

Fig. 14.3. Six-position optical switch with optical remote control. (a) Cross-section of the bundle of fibres: the central fibre (0) is permanently connected at a wavelength of λ_2 to fibre (7) for remote control of switching. Also, the central fibre (0) is connected at a wavelength λ_1 to one of the fibres $1, 2, 3, 4, 5$ or 6, according to choice. (b) Optical components include the bundle of fibres, a cylindrical lens, a surface with plane parallel faces which is a dichroic mirror reflecting λ_2 and a rotating mirror orientated by a step motor and reflecting λ_1. The rotating mirror and the dichroic surface are not perpendicular to $0z$ but inclined, the first to ensure propagation between fibre (0) and one of the six adjacent fibres, the second between fibre (0) and fibre (7). The length of the cylindrical lens is $\Lambda/4$, so that the output beam of light towards the mirrors is composed of parallel rays. The operator is connected to the switch by fibre (0), and this operates the motor which turns 60 degrees in azimuth on each step, as required, using λ_2. He is thus connected to the fibre of his choice to receive the required channel using λ_1

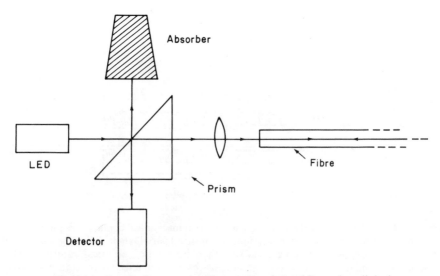

Fig. 14.4. Coupler/separator of beams permitting bilateral muliplexing

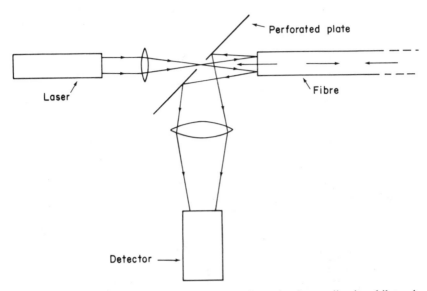

Fig. 14.5. Focal point coupler. This is a perforated mirror allowing bilateral multiplexing. Coherent emission from the laser is finely focused on to a plate which is marked and then perforated by a photolithographic process. The other face of the plate is a mirror

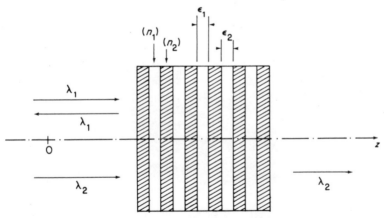

Fig. 14.6. Diagram of the principle of a dichroic mirror. A series of plates with plane parallel faces of refractive index n_1 and thickness ε_1 are inserted into a medium of refractive index n_2 and separated from each other by ε_2. For a light wave propagating along $0z$, the phase shift between two consecutive interfaces is $\Delta\varphi = (2\pi/\lambda_0)\varepsilon n$. It is 2π if $\varepsilon_1 n_1 = \varepsilon_2 n_2 = K \cdot \lambda_0$ with integer K. It is clearly possible to choose values of the parameters so that a wave of length $(\lambda_0)_1$ will be reflected almost totally, by addition of partial reflections, such that one or several others $(\lambda_0)_2, (\lambda_0)_3$ are transmitted. The rejection ratio can reach 29 dB in normal incidence. In oblique incidence, the performance is satisfactory up to about 20 degrees

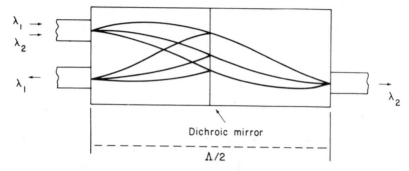

Fig. 14.7. Coupler/demultiplexer using a $2 \times \Lambda/2$ cylindrical lens and a dichroic mirror. Reciprocity allows use of the same component as either multiplexer or demultiplexer

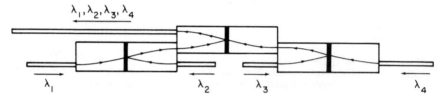

Fig. 14.8. Four-wavelength coupler/multiplexer

15　Large systems

"Man, this unknown"
Alexis Carrel (Nobel 1912)

(1)　The Biarritz experiment

15.1 EVOLUTION OF TELECOMMUNICATIONS IN FRANCE

The development of French telecommunications, conducted by the Minister of PTT, has allowed a rapid increase of the national telephone system, which has expanded from 7 million subscriber lines in 1975 to 20 million in 1982. This development continues and allows users to be offered new services which employ the latest technology. They depend on the general telecommunications directorate (DGT), the national centre for telecommunication studies (CNET) and the research laboratories of French industry.

Optical fibres and cables of optical fibres, generally called 'optical cables', allow transmission of much more information than metallic ones. It is certainly very interesting, technically and economically, to exploit this new technology, which arises in the accelerated evolution of the French national network, by assimilating temporal switching and digitization of signals. In this way, the French network can closely follow optics in its evolution.

Studies are oriented towards the establishment of a video-communication network by distinguishing two fields which are absolutely fundamental for this:

(1) The local network of subscribers; and
(2) Long-distance terrestrial and submarine transmission.

All must use new optical technologies. The local subscriber network is structured as a star to allow development of interactive services and it must be a very low-cost system. In this network signal management functions are performed electronically and the more substantial functions are ensured by the use of optical channels. These are transmission, switching and multiplexing in a tree network which consists of successive branches starting from a common trunk.

Long-distance transmission requires an increase in capacity of the inter-urban network of at least two orders of magnitude with a delay which is difficult to determine.* The solution is to use monomode fibres which are already economic for medium-capacity systems (140 Mbit/s) and capable of attaining throughputs of 100 (Gbit/s) × km and more by multiplexing carrier wavelengths at the extremities of the fibre. Modification of the infrastructure is not required and direct switching of light betweeen fibres can be envisaged.† The ambitious objective is to connect 6 million centres or organizations by 1992.

The essential advantage of temporal switching is to offer the possibility of constructing entirely digital networks, which leads to the attractive idea of an integrated digital network. Such a network would permit a genuine continuity of technology between the computer which processes and stores and the communication system which addresses and transmits.

At the beginning of the twenty-first century a digital network with integrated telecontrol and video-communication services can be envisaged. The first stage of an entirely digital network will occur, without doubt, with the commissioning, around 1987, of an integrated telephone and data service network, characterized by digitization of all transmitted signals and intended initially for professional use.

15.2 THE NECESSITY OF EXPERIENCE

It is clear that distribution on a national network of optical cables of one or two dozen television programmes (national and foreign channels) together with video recordings selected on demand by the user does not pose a difficult problem with present optical fibre and opto-electronic technology. In contrast, simultaneous interconnection on demand at peak hours of perhaps one or two million pairs of subscribers anywhere in France for sound and vision (vision-phone) communication represents the construction of an enormous system.

At present, the number of subscriber lines in the French telephone system is greater than 20 million. At peak hours 2 or 3 million French people speak to each other on the telephone, on the PTT network. How many of them would want to use a visionphone? Which visionphone service would they prefer? The answers to these questions are essential to provide correct development of telecommunications in France.

All projects and administrative plans are clearly subordinate to final acceptance by the customers of the service which is offerred to them. The reader will recall that in France and elsewhere there has been a temporary rejection of the telephone; in the 1950s it was claimed that the telephone was useless. Many individuals, businessmen and civil and military officials preferred a written note, message or telegram instead of a visit, that is, direct contact. In the USA there has been a setback, at least temporarily, of the 'visionphone', suggested by Bell Telephone to its customers.

* In these circumstances, a satellite such as Telecom 1 could be very useful.
† Switching is a prime necessity of the architecture of a network. In a monomode fibre system switching by directional coupler can be envisaged. Fabrication of an 8 × 8 matrix would seem possible.

It is certainly very difficult to foresee public reactions. Market research cannot establish the uncertain intentions of individuals, who do not know themselves what their reactions will be in the presence of events and situations which they cannot foresee.

Adaptation trials are necessary, that is, the specification of appropriate values of certain characteristics of the equipment and parameters of the system which make it more or less pleasant and useful to use. The objective is achieved when the user refuses all new modifications and, above all, a return to the previous system, that is, to lose a system to which he or she is accustomed to return to an old one which seems outdated. It is therefore necessary to experiment, to evaluate the needs, to establish the social effects and to master the industrial and economic effects.

The French Government's development plan for cabled videocommunication estimates that in France the demand for connection will reach 1.4 million within three years.

15.3 GENERAL OBJECTIVES OF THE BIARRITZ EXPERIMENT[5]

Biarritz, in the Atlantic Pyrenees *departement*, is a town of approximately 27 600 inhabitants on the Atlantic coast with a mean altitude of 40 m; it is 8 km from Bayonne and 183 km from Bordeaux. It has a railway station, an airport (Biarritz–Bayonne–Anglet) a municipal casino and two beaches. It is a resort and hydrothermal spa.

The Biarritz experiment should achieve three objectives:

(1) To master the concept, realization, exploitation and maintenance of a network of fibre optic cables on the scale of a town;
(2) To measure the attraction to the public of services for audiovisual exchange;
(3) To instigate development of a French optical communication industry, of which the Biarritz network will be the international shop window.
(4) Finally, as has been said above to test concepts for future networks.

At Biarritz, the principal concern will be to assess public attraction for a visionphone service and to observe choices.

15.4 THE EXTENT OF THE EXPERIMENT

The town of Biarritz has 19 600 houses distributed through eight regions. The objective is to connect 5000 subscribers in three stages from 31 March 1983 to 31 December 1985. (Connection of 30% of houses is anticipated.) The first stage to be put into operation involves a sheltered area in which direct reception of VHF and UHF television is difficult. This stage supports the connection of 1500 subscribers in the first two years. However, the buildings and cable network are dimensioned to allow connection of 2500 subscribers. Consequently, this first stage will use 10 000 km of optical fibres. This will be the first multiservice optical network in France, with a wide bandwidth, in contrast to an ordinary

162

telephone network with a narrow bandwidth. Transmission of sound and vision will be involved; more precisely, it will be possible to transmit to, or from, the subscriber:

(1) Two different television channels, chosen from 15 (the system can be extended to 30);
(2) One very high-quality stereo radio channel, chosen from 12;
(3) By the same terminal;
    (a) Visionphone in the Biarritz area;
    (b) Access to the worldwide network of telephone, facsimile, etc. services.

In a second phase, service companies will offer other facilities on an experimental basis, such as access to collections of pictures and catalogues, cinema according to choice, etc.* The present telephone network will not be modified and will continue to operate in parallel during the experiment.

15.5 GENERAL ORGANIZATION OF THE SYSTEM (Figs 15.1 and 15.2)

The system includes:

(1) A principal centre with;
    (a) The core of the network which carries out switching of the signals and their transmission to the secondary centres;
    (b) Broadcast management which allows control of the quality of programmes distributed to subscribers and connection of several sources of sound and pictures from near and far.
(2) Secondary centres, to which the subscribers are connected and which carry out;

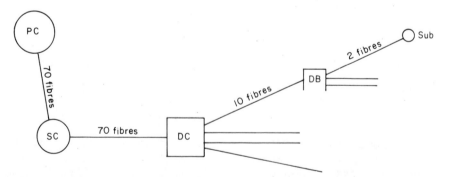

Fig. 15.1. Diagram of the cable distribution between the principal centre and the subscriber in the town of Biarritz. The longest length between centres is 2300 m and the longest link from a secondary centre to a subscriber is 1650 m. PC: principal centre; SC: secondary centre; DC: division centre; DB: distribution box; Sub: subscriber

* It is probable that interactive services will allow the number of television channels distributed to all subscribers to be limited.

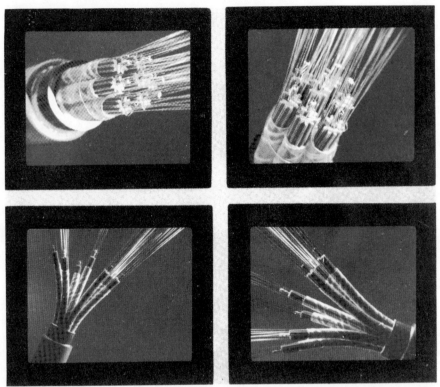

Fig. 15.2. 'Rush' optical cable. The basic element is the 'rush' of 10 optical fibres and the protected fibre resides in the helical groove of the 'rush'

   (a) Integration of visionphone traffic;
   (b) Selection of distributed programmes;
(3) A VHF–UHF receiving station which includes a network of antennas with receiving and control equipment necessary for the distribution of programmes by cable;
(4) A network of optical cables which includes;
   (a) A sub-network connecting the principal centre to the VHF–UHF receiving station and the secondary centres;
   (b) A distribution sub-network connecting the subscribers to the secondary centres.

15.6 THE SUBSCRIBER'S INSTALLATION (Fig. 15.3)

The subscriber's installation is connected to the network by one cable of two fibres, one for transmission and another for reception. This cable is connected to a control box which is both an optical/electronic interface and a switch which routes the electrical signals to the different equipment such as visionphone, television and high-fidelity sound channels. The complete terminal contains:

164

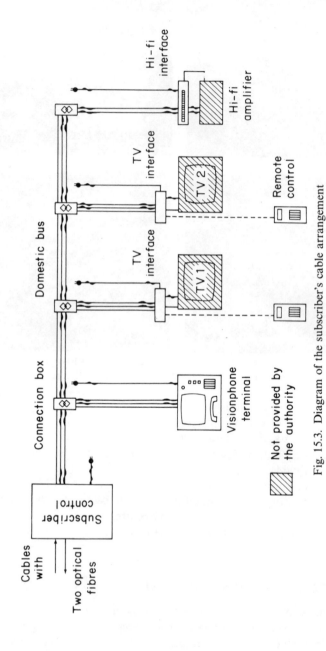

Fig. 15.3. Diagram of the subscriber's cable arrangement

Fig. 15.4. Visionphone terminal in use on the Biarritz experimental network

(1) The visonphone (hire of the equipment is included in the visionphone rental). This allows visionphone communication and also ordinary telephone communication and telecontrol service; later it will allow access to a range of programmes. (Fig. 15.4). The visionphone is equipped with keyboards for visionphone, ordinary telephone and videotext.
(2) Television receivers. These are not provided by the PTT administration. They are connected to the electrical network either by the antenna socket or by a peritelevision socket. An infra-red remote control box provided by PTT allows channel selection. The quality of the television image is superior to that which would be obtained with connection to a conventional external aerial. Furthermore, for reception of foreign programmes it is no longer necessary to have a multistandard receiver.
(3) A high-fidelity amplifier. This is not provided by the administration. A movable box provided by the administration allows a choice of programmes. The subscriber has at least as many universal sockets as there are terminals and these sockets are equally suitable for connection of the visionphone, television or high-fidelity channel. All the equipment can be interchanged between the sockets without modification to the cabling.

Fig. 15.5. New services. (a) Examination of distributed services for very good quality of pictures and sound; (b) access service, information research; (c) reception of a video library programme; (d) a transaction (reservation of a place on a high-speed train)

15.7 THE RANGE OF NEW SERVICES (Fig. 15.5)

The network installed in Biarritz allows two classes of service to be offered to subscribers:

(1) Distributed services; and
(2) Switched services.

15.7.1 Distributed services

Fifteen television channels and 12 Hi-Fi stereo channels will be available. All subscribers will be able to receive Spanish and various foreign channels on demand in addition to programmes of the French public television service. Video library programmes which offer subscribers a wide variety of titles will be added to these channels. Pricing will be a function of the service chosen.

15.7.2 Switched services

These permit the user to establish sound or vision communications on demand. The visionphone adds the caller's image on a screen to the telephone, and thus allows him or her to be not only heard but seen. Callers exchange their images only if they both wish to, otherwise, communication remains purely telephonic. Since the visionphone supports vision, it enables documents to be read at a distance, and modification to the visionphone camera position is provided for this purpose.

In rooms where lighting conditions do not normally permit viewing in colour a black and white camera is used. In contrast, a colour camera is used in professional premises (such as offices and banks) where the lighting can be specially arranged for colour images. The subscriber who has a personal video camera can connect it to his or her visionphone terminal.

All visionphones are equipped with a colour tube to receive images in colour or black and white. The visionphone can be used as an ordinary telephone for urban, inter-urban or international communication. Sound amplification, abbreviated numbering, automatic recall, etc. are all included in the equipment. In visionphone mode, communication will be charged at approximately double the simple telephone price.

15.7.3 Access services

Videotext allows stored information (such as reservations, numbers, operating instructions, complaints, weather forecasts, etc.) to be consulted by communication with a computer and, if necessary, to make a transaction such as a seat reservation.

In the second stage of the experiment it is anticipated that the user will be offered enhanced videotext services with not only graphic figures and tables composed of lines and alphanumeric characters but also true television type images and sound.

The list of possibilities offered to subscribers at Biarritz is unlimited, since the network of optical cables allows new services to be included in accordance with demand and availability without problems of complexity.

15.8 CONCLUSIONS

15.8.1 The Biarritz experiment in operation

Without doubt, the Biarritz experiment confirms the public's preference for an image. It is difficult to grasp this power of an image, even a stationary one. We certainly analyse the power of sounds better than that of images which initiate complicated and little-understood reactions in us.

How is it that, with the same poor definition of present colour television, we much prefer, for example, a landscape painted by Cézanne to the actual landscape presented on the same small screen? By what transformation of the signals can one change from the actual landscape to the landscape seen by Cézanne and successfully transmit it with the same definition? Where does the power of the image go?

If the result of the Biarritz experiment on the attraction of videocommunication is positive, construction of new switched telecommunication networks with correspondingly wide bandwidths will be agreed.

15.8.2 Towards entirely optical systems

Putting recent events into perspective allows an estimate of the number and significance of events in the near future to be made. The fact that the field of optics is developing rapidly and that optical communication technology is still very far from its limits must be kept in mind. Each year, new techniques are added to those which are already available, and new areas with vast horizons are opened to us. A high level of ambition is permitted.

However, it is necessary to create, that is, to invest, now. The key to success lies inevitably with the development of new wide-bandwidth services.

(2) Submarine cables

15.9 WORLD TELECOMMUNICATIONS AND SUBMARINE CABLES

To transmit telecommunications signals over long distances, account must be taken of the curvature of the earth. There are two main practical methods for this:

(1) A satellite relay; or
(2) A cable.

Very unspectacular when it settles discreetly into obscurity at a great depth, to stay there in conditions of perfect tranquility, the submarine cable is consequently the most unappreciated of intercontinental transmission systems. However, cables already carry more than half the world's international communications, and the tendency is for them to increase. With the advance of optical fibres, a considerable growth in cables can be forseen.

Transmission of signals by submarine cable is a quite economic enterprise. Four countries have an industry of this kind—the USA, the UK and Japan, which are islands, and France.* This industry requires:

(1) A means of fabrication of long unit lengths of cable which must be very reliable, able to withstand pressure at great depths and tension during laying

---

* The USSR produces its own submarine system for internal use.

and re-laying (for example, occasions when a length of 15–20 km of cable can be suspended from the cable-laying ship in a heavy sea);

(2) A means of fabricating submerged repeaters and coastal terminal equipment;

(3) A means of transporting and laying from several hundred to several thousand kilometers of cable using a specialized ship (for example, with which to lay cable in the Atlantic in one, two or three passes);

(4) A knowledge of techniques for laying and recovering the end of a cable by dredging at great depth and splicing at sea;

(5) A secondary burying technique to permit cable installation on continental plateaus at depths to around 600 metres; and

(6) Experience and practical knowledge.

Reliability is an essential characteristic of marine and submarine systems. A fault on a cable in mid-ocean requires long and costly work and involves a loss of important traffic.

Modern submarine systems are produced to operate at full capacity for 25 years and, if necessary, at reduced capacity for up to 32 years. During such a period of time a terrestrial system requires several hundred maintenance and repair operations.

From 1985 onwards civil fibre optic submarine systems will be installed throughout the world. Military systems and sub-systems have already existed for several years.

In the ocean there is a large network of abandoned submarine telegraph cables and more than 265 000 km of coaxial cable in service. (The UIT [Union Internationale des Télécommunications] publishes a list of cables forming the world submarine network.)

France is particularly interested in the seas which border it—the English Channel, the North Atlantic and the Mediterranean. The CGE group (Compagnie générale d'électricité) has created a division for submarine telecommunication systems—SUBMARCOM. Production of the cables is the responsibility of the Lyon Cable company at Calais and the terminal equipment and the repeaters are the responsibility of CIT Alcatel at Villarceaux. In 1979 CGE had produced 20 000 nautical miles* of submarine links and 1500 submerged repeaters.

Table 15.1 indicates (errors and omissions excepted) the principal cables (of more than 2000 km) laid during the last ten years with at least one French participation.

In France submarine cables carry the following administrative services:

(1) General direction of telecommunications;

(2) Direction of international industrial affairs (with the help of CNET);

(3) Direction of the external telecommunications network (preparation and execution of projects, maintenance of the network and management of the cable fleet).

---

* A nautical mile is 1852 m.

Table 15.1

Name of cable	Constructor	Capacity (telephone circuits)	Length (nautical miles)
Ariane (France–Crête)	CGE	640	1336
TAT-6	STC-CGE-Bell	4000	3396
Antineá(Morocco–Senegal)	CGE	640	1458
Fraternité			
(Senegal–Ivory Coast)	CGE	640	1415
TAT-7	STC-CGE-Bell	4200	3400

In the UK STC (Standard Telephones and Cables Limited), an associate of ITT, is, without doubt, the largest manufacturer, with 80 000 nautical miles and 5800 repeaters. The British company 'Cable and Wireless Ltd' is an international group which owns, or has interests in, 32 000 nautical miles of cable throughout the world.

15.10 COMPLEMENTARITY OF SUBMARINE CABLES AND SATELLITES

Television networks and the press have widely reported successful launchings, putting into orbit and commissioning of telecommunications satellites. Consequently the public believes that the satellite, a new and expanding technique, is the only approach, and cable, an old and outdated method, is to be abandoned. This is a mistake.

In fact, the satellite is a relay for radio propagation through open space by crossing the Earth's atmosphere twice. The great advantage of this relay is that it covers a very large geographical area which, for a geostationary satellite, represents approximately three-quarters of a hemisphere. It is therefore suitable for links with mobile and isolated terrestrial stations. The satellite has advantages if the distance is great and the volume of traffic is small.

However, the capacity of not only one but all geostationary satellites is limited. There is a moving curved line which surrounds the Earth at an altitude of around 36 000 km where it is possible to place an object temporarily in a geostationary position by giving it a suitable velocity. Satellites must be sufficiently separated so that they do not obstruct each other and they must share the frequency band of radio transparency of the atmosphere at around 10^{11} Hz, which is used for microwave links.* In contrast, use of submarine cables to establish fixed point-to-point links is unlimited and each silica fibre, in the best optical band currently used, could theoretically handle some 10^9 Hz. This is a substantial speech capacity.

---

* Part of the orbit covering Canada and the USA is already saturated for the 4 and 6 GHz bands

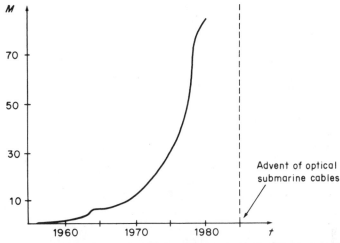

Fig. 15.6. The great increase in submarine cables. *M* is the product of the number of millions of telephone circuits and the number of nautical miles covered

It is important to remember that the capacity of satellites is limited, and that their lifetime depends on the propellant with which they are provided; on the other hand, the capacity of an optical guide is considerable and the number of cables which can be laid is unlimited.

On 28 June 1965 the Early Bird satellite entered commercial use with a capacity of 240 telephone channels. A little later, the ATT company dismantled its submarine services and sold its cable factory!

However, development of world telecommunications continued and increased.

Table 15.2. The increase in capacity and decrease in price of telephone channels

Submarine cable:	The capacity is multiplied by	40 in 10 yr
(over 25 years)	The price per circuit is divided by	6 in 10 yr
Satellites:	The capacity is multiplied by	33 in 10 yr
(over 15 years)	The price per circuit is divided by	4 in 10 yr

New optical device will intervene largely in favour of submarine cables

Table 15.3. French intercontinental circuits

	1974	1980
Submarine cables	595	2757
Satellites	363	1566
Ordinary radio channel	87	38
Total	1045	4361

(Multiplied by 4 in 6 years)

New optical and submarine cable techniques will increasingly be used (Fig. 15.6: see also Tables 15.2 and 15.3).*

15.11 SUBMARINE CABLES BETWEEN EUROPE AND THE USA

For twenty years the demand for trans-atlantic communication has increased by 10–25% per year, and it doubles in less than five years.

Table 15.4 summarizes the evolution of the system capacity. At present, the use of three different INTELSAT satellites and five metallic submarine cables TAT 3–TAT 7 provides good diversity of transmission. The European administration and the American telecommunication companies have discussed terms for traffic sharing between the new generation of satellites INTELSAT 6 and future submarine optical cables TAT 8 and TAT 9.

(*Note*: The effective propagation time of the signal by geostationary satellite is 500 ms; by optical cable it will probably be less than 50 ms taking account of regenerators and at least 33 ms.)

The principal characteristics of the new services introduced to facilitate communication have been analysed:

(1) The diversity of information rate from 2.4 kbit/s for text processing to 2 Mbit/s for a vision conference.
(2) The diversity of transmission times from several seconds for a telescopy to an hour or more for a vision conference.
(3) The variation of traffic density during a year, a week and a 24-h day.
(4) The speed of traffic through a geographical area.

15.12 THE NEXT TAT TRANSATLANTIC CABLES

Finally, it has been decided that the optical cable TAT 8 will be put into service in June 1988. This will be produced in three geographical sections, forming a

Table 15.4

Year	Designation of cable	Number of circuits	Cumulative total	Diameter of cable (cm)	Separation of repeaters (km)
1956	TAT1	36	36	1.57	70.5
1959	TAT2	36	72	1.57	70.5
1963	TAT3	140	212	2.54	37.1
1965	TAT4	140	352	2.54	37.1
1970	TAT5	820	1 172	3.81	18.6
1976	TAT6	4 200	5 372	4.32	9.5
1983	TAT7	4 200	9 500	4.32	9.5
(1988)	TAT8	(7 680)	(17 180)	(2.5)	(45)

* Without doubt, the day will come when telecommunications satellites also use optical techniques, at least betweeen each other.

Y in the ocean. The American responsibility will cover the principal branch of the Y, from Tuckerton (USA) to a point near to the European coast. The French responsibility will extend from this point to Penmach (Finisterre). The British responsibility extends from this same point in the Atlantic to the English coast.

The American sub-system, entrusted to Bell, consists of a cable of four optical fibres. The French and British sub-systems, made by CGE (Submarcom) and STC, respectively, each consists of a cable of two fibres. The three sub-systems will be compatible and joined by a submerged black box which connects them and includes remote control. Thus the system will permit not only transmission between Europe and the USA but also between France and the UK. The features of the system are:

(1) Silica monomode optical, fibres,
(2) Pulse code modulation (PCM) with a single format for telephone, video and data;
(3) A gross capacity of 7680 telephone channels or their equivalent, that is, a rate of 274 Mbit/s;
(4) Transmission in a digital format which permits the use of active traffic concentrators such as 'CELTIC'. This French concentrator is used for the whole system and multiplies the telephone channel capacity by six, enabling around 40 000 effective channels to be carried.

15.13 THE FRENCH SYSTEM 'S280' (Fig. 15.7)

The technical characteristics of the US, UK and French sub-systems of TAT 8 will certainly be similar. The French administration of PTT has developed the S 280 system, of which the French sub-system of TAT 8 will be an application. The first commercial realization of this system will be the submarine optical link between Marseilles and Ajaccio in 1985.

In the standard version, the S 280* system uses two pairs of monomode fibres, each carrying two 140 Mbit/s pulse trains in one direction, which makes four 140 Mbit/s trains in each direction. The maximum capacity is 7500 km and the maximum laying depth is 6500 m. The repeaters are bidirectional and separated by at least 45 km; each contains two pairs of regenerators. The silica fibre in the cable has an attenuation of 0.53 dB/km at 1.3 μm.

The transmission capacity for two pairs of fibres is 7680 telephone channels in each direction, or their equivalent. Traditional telephone channels are at 64 kbit/s and Television signals at 34 Mbit/s.

15.13.1 The CELTIC concentrator

Transmission in digital format allows the use of active traffic concentrators. The French concentrator uses the inactive times of circuits during normal operation, without a concentrator. Concentration is based on three principles:

---

* 280 Mbit/s on one fibre.

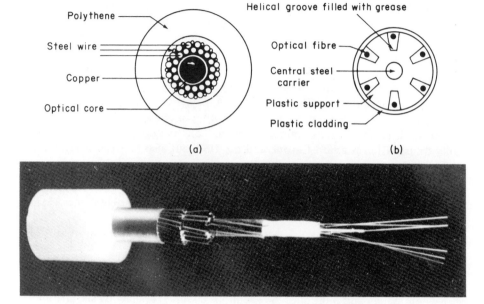

Fig. 15.7. Cable for system S280. (a) Section of cable; (b) Section of optical core; (c) deep-water submarine cable containing three pairs of fibres

(1) The technique of interpolation of speech during silences in a conversation;
(2) Adaptive linear prediction which reduces the description of the speech signal;
(3) Coding the signal by digital words at a variable rate.

These techniques, used simultaneously, reduce the traffic in a ratio of 6:8, which allows 180–240 conversations to be transmitted using 2 Mbit/s instead of 30. Data transmission is performed without concentration.

15.13.2 The principal characteristics of the cable

The normal point-to-point link uses two pairs of fibres. This can be reduced to one pair or increased to three. They can be separated, at sea, into two cables to serve two different landing points:

(1) Diameter in air: 25mm;
(2) Breaking strain: 140 kN;
(3) The fibres are placed in a watertight optical core, accommodated in helical grooves in a plastic retainer with a slack of 1.1%. The helical grooves are filled with waterproof grease and regularly spaced watertight bushes prevent longitudinal propagation of water after an accidental rupture;
(4) Copper conductors provide a power supply for the repeaters;
(5) A steel covering ensures both resistance to tension and protection of the optical core from pressure.

15.13.3 Regenerating repeaters

Each repeater detects the optical signal with a photodiode and amplifies it with automatic gain control. The signal is returned to standard form, resynchronized and re-emitted by a laser diode. A redundancy of three spare diodes is provided with switching by remote control. Monitoring circuits transmit the necessary information to terminal stations and carry out the commands, which have been standardized for the French digital network and allow computer-aided operation.

15.13.4 Economic analysis

This shows that with heavy traffic over long distances such as the North Atlantic communication by 'S280' optical cable will cost three times less than by satellite.

15.13.5 Critical remarks

The choice of wavelength $\lambda_0 = 1.3\,\mu\text{m}$ is currently the best for silica and laser sources using InP. Avalanche photodiodes are not used on account of the complicated peripheral circuits required.

Unfortunately, the monomode fibre will be very underused at 3.2 GHz × km when it offers almost 40 GHz × km. The bit rate of 280 Mbit/s is limited by the slow response time of terminal components (semiconductor laser diodes and PIN diodes).

Multiplexing of several wavelengths on the same fibre is not used because the reliability of the components is still uncertain.

The cable of six fibres, with its copper conductors which are necessary to supply power to the repeaters, has a volume approximately six times smaller than that of a coaxial cable of the same capacity.

15.14 THE FRENCH EXTERNAL NETWORK IN EUROPE AND THE MEDITERRANEAN

The French administration forecast an installation in 1985 of more than 500 telephone circuits by satellite under the EUTELSAT organization in order to make the network secure. Furthermore, in parallel with the increase in telephone and television traffic in Europe, the development of new services has led the administration to instal a satellite system TELCOM 1 for France and Western Europe. However, in the Mediterranean basin, which constitues a special case, priority has been given to submarine cable.

16 Future prospects for optical cable systems

In this chapter we shall attempt to put the progress of system techniques for optical cable transmission during the next fifteen years into perspective without claiming to predict economic and social effects which depend upon excessively uncertain human factors.

16.1 OPTICAL GUIDES, MATERIALS AND WAVELENGTHS

The most important component of the system is certainly the fibre, a marvellous result of glass drawing at high temperature. All the systems which we have mentioned, and would wish to consider, are based on optical fibres.

The three enemies of transmission are dissipation, divergence and dispersion. The first two have already been largely mastered. The third remains—the lengthening as a function of distance of the shaft of light which forms the group of photons of the 'signal'.

Linear electromagnetic theory and laboratory measurements on silicon show that a strand of this material 1 km long used with a single low-power source of monochromatic light has a transmission capacity limited to 600 or 800 Gbit/s, which is nearly a terabit per second. Unfortunately, irregularities of the interface of present monomode fibres give a practical effective capacity of several tens of Gbit/s over a kilometre distance.

However, several light sources may be used simultaneously with the same fibre, thereby multiplying the information rate, provided that the sum of the powers propagated together does not exceed a certain limit.

Advancing into the near infra-red, towards longer wavelengths, the attenuation of optical power per kilometre is reduced. For the same good-quality silica fibre in a cable one finds:

At λ_0 (nm)	850	1200	1550	2000
A (dB/km)	2.1	0.54	0.31	(0.19)

Table 16.1

Year	1985	1990	2000
Operating wavelength (nm)	850	850	(850)
	1200	1200	1300
	1300	1300	1550
		1550	2000
			2500

Table 16.2

Material attenuation per kilometre	Present results		Expected results	
	A(dB/km)	λ_0(nm)	A(dB/km)	λ_0(nm)
Vitreous SiO_2	0.2	1 500	0.2	1 500
Vitreous GeO_2	13	2 500	0.1	2 500
Heavy metal glasses				
PbO, Tl_2O, Bi_2O_3			0.02	3 000
Fluoride glasses[a]				
CaF_3, BaF_3	20	2 500	0.001	3 500
Polycrystaline materials				
KRS5, TlBr	400	10 600	0.001	7 000

[a] $Zr F_4$: expected result 0.001 dB/km at 2550 nm.

Hence, range is increasing. However, the power threshold necessary for detection tends to increase with wavelength, and the problem of dispersion above 1300 nm becomes more difficult to resolve. However, it will be resolved, and the separation of repeaters will be doubled or quadrupled during the next few years, so that 100–200 km can readily be achieved with silica fibres. The operating wavelengths will also evolve, particularly for long distances and large cable systems in order to avoid regenerating repeaters which must be supplied with power at a distance (Table 16.1).

The market is currently dominated by synthetic vitreous silica, and it appears that this situation is stable. However, other materials have already been proposed and will appear on the market according to their price and performance (Table 16.2).

16.2 LIGHT SOURCES AND DETECTORS: THE PRINCIPAL COMPONENTS

The first generation of components in the 800–900 nm band has largely been developed and is currently used for distances less than 10 km. A certain lack of integration is to be regretted. Small complete systems are being developed and marketed for signal transmission over less than 1 km for which installation becomes economic when the distance to be covered exceeds a few tens of metres.

Table 16.3. Power and mean lifetime of light sources

Year	1985		1990		2000	
Light source	Power (W)	Lifetime (h)	Power (W)	Lifetime (h)	Power (W)	Lifetime (h)
Light-emitting diode	0.1	6.10^4	0.2	7.10^4	0.5	8.10^4
DH laser diode	0.6	5.10^4	1	1.10^5	2	2.10^5
DH laser diode (integral fibre)	1	5.10^4	2	—	3	—

The second generation of components in the 960–1660 nm band provides light souces of very good reliability over several years and InGaAs on InP light detectors which, at present in the laboratory, have response times limited to 30 ps; this corresponds to detection of a data rate of 13 Gbit/s.

Ternary compounds (III–V) $In_{0.53}Ga_{0.47}As$ allow fabrication of PIN photodiodes and associated amplifiers with uniform sensitivity in the 960–1660 nm band, a quantum efficiency of 70% and results comparable with those of (II–VI) compounds $Hg_{0.3}Cd_{0.7}Te$ at 1300 nm.

Emission of two light sources from the same LED and reception with the same photodiode can be foreseen by using separate modulation techniques—for example, digital for one and analogue for the other. A third generation will appear in the 2000–2500 nm band. (Table 16.3). Other photodetectors are being actively researched in association with low-noise amplifiers for reception of longer wavelengths, and several systems are already available which give satisfactory results.

16.3 LONG DISTANCES AND WIDE BANDWIDTHS

As a consequence of technical progress, which has reduced attenuation per kilometre, throughput is limited by the required bandwidth and not the threshold of optical power necessary for detection. This situation can lead to the simultaneous use of several light sources which share the total required bandwidth. Multiplexing of light souces is developing, using dichroic mirrors or filters, diffraction gratings and other devices. This is economic only if the price of the coupling and multiplexing/demultiplexing equipment at the terminals is less than that of fibres to be added to the cable to give the same result.

For very large throughputs over several hundreds of kilometres it will probably be necessary to increase the data rate of the fibre. The solution envisaged is to propagate solitons of light which retain their property of coherence by a non-linear effect, which is possible over a long distance only if the kilometric attenuation is very much reduced (for example, to around 0.001 dB/km.). Reduced attenuation is, therefore, still important (Table 16.4).

Table 16.4. Bit rate $(Gbit/s) \times km$ of linear region of silica fibre

Year	1985	1990	2000
Multimode SI	0.06	0.1	0.2
Multimode GI	5	8	10
Monomode	20	50	70

16.4 NEW DIRECTIONS OF RESEARCH

16.4.1 New materials, new glass technologies

The covalent chemical bond produces an excellent cohesion, and it is this which produces the remarkable properties of silica, which amount to an inorganic refractory ceramic.

Others can be produced, at low temperature, such as vitreous aluminium phosphate $AlPO_4$. This substance has a compact crystalline structure very similar to that of silica. However, it is impossible to produce it by fusion, since it decomposes before it melts.

To produce it, a crystaline precipitate of $AlPO_4$; HCl; $(C_2H_5OH)_4$ is prepared by reacting aluminium trichloride and phosphoric acid in ethyl alcohol at sub-zero temperatures. The structure of the precipitate is of cublic form. On heating gently to 100 °C the cubes link together to form a vitreous three-dimensional lattice of aluminium, oxygen and phosphorus which is refractory up to 1600 °C. These low-temperature glasses are produced, among others, at the Institute of Technology in Tokyo. They have a very high transparency, do not contain irregularities due to mixing of a diameter greater than 100 Å and could allow rapid fabrication of optical ribbons.

16.4.2 Control of the spectral line emitted by a semiconductor laser: digital frequency modulation

The spectral purity of injection lasers can be improved by means of a distributed feedback device. The active layer of a semiconductor laser includes a periodic diffraction grating whose effect is added to that of the Fabry–Pérot cavity in order to select a single mode of resonance, that is, a single narrow emitted spectral line.

In May 1983 Bell Laboratories announced the C^3 laser, an arrangement of two aligned lasers, one behind the other, operating together on the same plane active layer. The two lasers have different lengths, and thus have only some modes in common within the limits imposed by diffraction. In fact, a single mode benefits from the maximum gain, and the probability that the mode of the emitted line changes randomly to a neighbouring mode is less than 10^{-10}.

By placing one of the diodes below the laser threshold and modifying the current which feeds the other diode a change from one modal line to another occurs, that is, the wavelength changes. Fifteen wavelengths in an interval of $300\,\text{Å}$ can be passed through in succession by making successive changes of several milliamperes in the pilot diode current.

In a monomode fibre at $\lambda_0 = 1.55\,\mu\text{m}$ and over a distance of $110\,\text{km}$ Bell Laboratories have obtained a data rate of 1 Gbit/s and 420 Mbit/s over 160 km with an error rate of 5.10^{-10}.

Other applications involve transmission on several wavelengths; for example:

(1) One for the '0' symbol and the other for the '1' symbol;
(2) Four wavelengths associated with the binary numbers $00, 01, 10$ and 11;
(3) Eight wavelengths associated with each of the three-bit binary numbers.

At the receiver, the wavelengths (2, 4 or 8) are separated by a diffraction grating.

16.4.3 Use of non-linear effects

Polarization $\vec{P}$ induced in a linear dielectric by an electric field $\vec{E}$ is given by:

$$\vec{P}_L = \chi^{(1)}\vec{E}$$

where $\chi^{(1)}$ is, in general, a tensor of rank 2.

Furthermore

$$\vec{D} = \varepsilon_0\vec{E} + \vec{P} \quad \text{and} \quad \vec{D} = \varepsilon^{(1)}\vec{E}$$

where $\varepsilon^{(1)}$ is also a tensor of rank 2.

The non-linear response of a medium can be expressed by a limited expansion of the following form:

$$P_{NL} = X_1 \cdot E + X_2 E^2 + X_3 E^3$$

For example:

$$\vec{P}_1 = \chi^{(1)}\vec{E}, \vec{P}_2 = \chi^{(2)}|E|\vec{E} \quad \text{and} \quad \vec{P}_3 = \chi^{(3)}|E|^2\vec{E}$$

The coefficients $\chi^{(n)}$ are tensors of order $(n+1)$, called 'non-linear susceptance'. The terms decrease very rapidly with increasing n.

A term of the P_2 type permits doubling of the frequency of a light source (for example, using a YAGNd laser at $1.064\,\mu\text{m}$ to radiate at $0.532\,\mu\text{m}$ in the blue–green). For reasons of symmetry, this term is zero in amorphous media, gases, liquids, glasses and plastics.

The P_3 term which, although very small, is not zero in glasses produces:

(1) Due to its real part, a non-linear phase index (automodulation of phase) and a degenerate mixing of four photons;
(2) Due to its imaginary part, Raman and Brillouin effects.

The composition of the wave vectors determines the forward direction with Raman scattering and the reverse direction with Brillouin scattering. The maximum effective Brillouin section in vitreous silica is about three hundred times that of Raman scattering.

Brillouin oscillations have been used to convert continuous laser light into 10 ns pulses by means of a fibre, but details of possible uses of these effects is outside the scope of this book. It is appropriate, however, to draw attention to the wealth of possibilities.

16.5 THE DEVELOPMENT OF OPTICAL SYSTEMS

At the end of 1983 ATT, the US telephone and telegraph company, achieved the commissioning of a network of more than 30 million km of optical telecommunication circuits. The superiority of an optical system over an electrical one can be appreciated from the following ratios:

(1) The transmission capacity of the optical system is 10 000 times greater. (This ratio is that of estimated physical limits.)
(2) The energy loss, in the course of propagation of a signal, is a hundred times less. (This ratio should improve.)
(3) The error rate is ten times less.

The solution to the problems of guided telecommunication is certainly optical, that is, one to be reached by the use of photons.

Principal symbols and abbreviations

A	coefficient; rectilinear attenuation (dB/km); area
A_c	area of the cross-section of the core
A_s	area of the source
Å	Angström (10^{-10} m)
a	radius of the cylindrical rod; radius of the core–cladding interface
APD	avalanche photodiode
B	bandwidth (Hz)
b	external radius of the cladding; binary digit rate (bit/s)
°C	degrees Celsius
c	velocity of light in free space (equation (2.3))
d	thickness; distance
DH	Double heterojunction
DR	depletion region.
$\vec{E}$	electric field vector
ENP	equivalent thermal noise power (equation (12.20))
F	function; normalized frequency (dimensionless, equation (4.4))
f	frequency (Hz)
GI	graded index
g	exponent of the 'power law' formula (equation (5.1))
$\vec{H}$	magnetic field vector
h	Plank's constant (equation (2.1))
$h(t)$	impulse response; instantaneous optical output power
$i(t)$	instantaneous current
$i_\phi(t)$	photocurrent
$i_{th}(t)$	thermal noise current
$i_q(t)$	quantum noise current
J_1	Bessel function of the first kind, order 1
j	imaginary unit
K_1	Bessel function of the second kind, modified, order 1
°K	degrees Kelvin (absolute temperature)
K	parameter proportional to the square of the signal to noise ratio, $K = Q^2$

k	Boltzmann's constant
$\vec{k}$	wave vector
k	wave number, modulus of k, $k = \omega/v$ (equation (2.21))
k_0	wave number in free space (equation (2.22))
L	length
L_c	critical length
l	length, distance
M	point; mean avalanche multiplication factor
m	avalanche multiplication factor; propagation mode
MDP	minimum detectable power
N	group refractive index
NA	numerical aperture
n	refractive index (of phase)
n_1	index along the $0z$ axis
0	centre, origin
$0z$	axis; general direction of propagation
P	power
P_e	error probability
P_0	optical power
PIN	simple semiconductor photodiode
$p(t)$	instantaneous optical power
Q	parameter (equation (12.3))
q	charge on the electron
r	radius, distance from the axis of revolution
r, ϕ, z	co-ordinates
S	size, extent
S_e	sensitivity (equations (10.4), (10.5))
SI	step index
sr	steradian
T	temperature; basic time interval or slot
$T_{\text{half-max}}$	duration or magnitude of received pulse, measured at half of the maximum optical power
v	variable; group velocity
V	potential
v	phase velocity
W	power
x, y, z	co-ordinates
α	attenuation coefficient
β	wave number, real part of $k = \beta - j\alpha$; wave number along $0z$: $\beta = k_z = 2\pi/\lambda_z$
Γ	spectral density
γ	electromagnetic radiation
Δ	Laplacian; relative refractive index
ε	electric permittivity (dielectric constant); ε_0 electric permittivity of free space ε_r relative electric permittivity (equation (2.14))

η	quantum efficiency
θ	angle of ray with $0z$ axis
θ_c	critical value of θ
θ_{ACC}	angle of acceptance
λ	wavelength
λ_0	wavelength in free space
λ_z	wavelength in the $0z$ direction
μ	magnetic permeability;
	μ_0 magnetic permeability of free space (equation (2.10)); μ radius parameter (integer)
ν	optical frequency; azimuth parameter (integer)
σ	conductivity
ϕ	azimuth
φ	phase
ψ	wave function $\psi(z,t), \psi(r,\phi,z,t)$
Ω	solid angle
Ω_{ACC}	solid acceptance angle
ω	angular frequency (pulsatance) (equation (2.20))

References

1 *Revue technique Thomson CSF* **6**, No. 4, December, Gauthier-Villars, Paris, 1974.
2 M. K. Barnoski, *Fundamentals of Optical Fiber Communications*, Academic Press, New York, 1976.
3 J. A. Arnaud, *Beam and Fiber Optics*, Academic Press, New York, 1976.
4 S. E. Miller and A. G. Chynoweth, *Optical Fiber Communications*, Academic Press, New York, 1979.
5 A. Cozannet, J. Fleuret, H. Maitre and M. Rousseau, *Optique et télécommunications*, Collection CNET-ENST, Eyrolles, Paris, 1983.
6 M. Papuchon and C. Puech, Integrated optics a possible solution for the fibre gyroscope. *SPIE* **157**, 218–24, Laser inertial rotation sensors, 1978.
7 J. A. Bucaro and J. H. Cole, *Acousto-optic Sensor Development*, Naval Research 4 Lab Washington, DC.
8 A. W. Snyder and J. D. Love, *Optical Wave Guide Theory*, Chapman and Hall, London, 1983.
9 J. M. Hammer, Optical waves guides modulation techniques. In *Fiber and Integrated Optics* D. B. Ostrowsky, Plenum Press, New York, 1978.
10 J. D. Montgomery and E. W. Dixon, Fiber optics to the year 2000. *Fiber and Integrated Optics*, **3**, No. 4, 1981.
11 J. C. Campbell, Photodetectors and compatible low-noise amplifiers for long wavelength light wave systems. *Fiber and Integrated Optics*, **5**, No. 1, 1984.
12 H. Hodara, Fiberoptic receiver performance, a tutorial review. *Fiber and Integrated Optics* **4** No. 3, 1983.
13 S. D. Personick, Receiver design for digital fiber optic communication systems. *BSTJ*, July-August, **50**, No. 1, 843–86, 1973.

Index

188